SpringerBriefs in Applied Sciences and Technology

SpringerBriefs present concise summaries of cutting-edge research and practical applications across a wide spectrum of fields. Featuring compact volumes of 50 to 125 pages, the series covers a range of content from professional to academic.

Typical publications can be:

- A timely report of state-of-the art methods
- An introduction to or a manual for the application of mathematical or computer techniques
- A bridge between new research results, as published in journal articles
- A snapshot of a hot or emerging topic
- An in-depth case study
- A presentation of core concepts that students must understand in order to make independent contributions

SpringerBriefs are characterized by fast, global electronic dissemination, standard publishing contracts, standardized manuscript preparation and formatting guidelines, and expedited production schedules.

On the one hand, **SpringerBriefs in Applied Sciences and Technology** are devoted to the publication of fundamentals and applications within the different classical engineering disciplines as well as in interdisciplinary fields that recently emerged between these areas. On the other hand, as the boundary separating fundamental research and applied technology is more and more dissolving, this series is particularly open to trans-disciplinary topics between fundamental science and engineering.

Indexed by EI-Compendex, SCOPUS and Springerlink.

Abdallah Hamed

Cascaded Interferometers and Their Medical Applications

 Springer

Abdallah Hamed
Department of Physics, Faculty of Science
Ain Shams University
Cairo, Egypt

ISSN 2191-530X   ISSN 2191-5318 (electronic)
SpringerBriefs in Applied Sciences and Technology
ISBN 978-3-031-64534-1   ISBN 978-3-031-64535-8 (eBook)
https://doi.org/10.1007/978-3-031-64535-8

This Springer imprint is published by the registered company Springer Nature Switzerland AG
The registered company address is: Gewerbestrasse 11, 6330 Cham, Switzerland

If disposing of this product, please recycle the paper.

To the spirit of my parents
To my family
To the Department of Physics, Faculty of
Science, Ain Shams University

Preface

Many objects examined under the microscope influence the phase but not the amplitude of the incident light, and the image is of such poor contrast that much structural detail is lost. Multiple beam interference, two-beam interference, and phase-contrast microscopy all present means of overcoming this difficulty. In multiple beam interference, an optical flat is matched against the specimen, and the resulting interferometer is suitably illuminated and viewed under a microscope. This procedure is based upon the established practice with Fabry–Perot interferometers, which was introduced by TOLANSKY (1948), who used metal-coated interferometer surfaces to enhance both the sensitivity and the scope of the method.

In this book, we suggested cascaded two-beam and multiple beam interference to further improve image contrast. In Chap. 1, modulated two-beam interference is considered using Fourier optics. We presented rectangular, triangular, and truncated Gaussian models for the fringe shift that occurred in the image. In Chap. 2, we suggested a cascaded Michelson interferometer. We calculated the refractive indices of microscopy images, e.g., glass fibers, using a modified Michelson interferometer. In Chap. 3, we investigate step-index fibers using multiple laser beam interferometers. In Chap. 4, modeling of the fringe shift for unclad glass fibers using ordinary multiple beam interference is described. The recognition of some modulated apertures using a cascaded Fabry–Perot interferometer is discussed in Chap. 5. In the next section, four chapters on medical applications are presented. Retinal artery image processing using higher orders of two-beam interference is investigated in Chap. 6. The investigation of kidney images using a cascaded Fabry–Perot interferometer is presented in Chap. 7. The image processing of coronavirus using interferometry was investigated in Chap. 8. Finally, the investigation of colon images using cascaded interferometry is outlined in Chap. 9.

Cairo, Egypt   Abdallah Hamed

Contents

Part I
Cascaded Interferometers

Chapter 1
Modulated Two-Beam Interference Using Fourier Optics

1.1 Introduction

The present chapter addresses a simulation process to compute the refractive index distribution assuming three different models of fringe shifts from two-beam interference [1]. In the following section, a theoretical analysis is presented considering monochromatic light for the illumination of the optical setup and considering the concept of Fourier optics [2–9]. In the following discussion, the results are presented, followed by a conclusion.

1.2 Theoretical Analysis

A Michelson arrangement illuminated with monochromatic light is used as a processor, as shown in Fig. 1.1. The examined phase object is placed in one of the arms of the interferometer, while an inclined plane wave, making an angle (α) to the normal serving as a carrier wave, is incident upon the second arm of the interferometer.

In the imaging plane of coordinates (x, y), the carrier wave is represented as follows:

$$A_c(x, y) = \beta_1 a(\lambda) \exp\left[-j\left(\frac{2\pi}{\lambda}\right) y \sin(\alpha)\right]; \ \beta_1 = r_1.r_s.t_s \qquad (1.1)$$

r_s and t_s are the amplitude reflection and transmission coefficients of the beam splitter (BS) respectively, and r_1 is the amplitude reflection coefficient of mirror M_1. $a(\lambda)$ is the amplitude spectral response of the light source.

In the former case of monochromatic illumination, the complex amplitude of the phase object received in the imaging plane is written as follows:

A. Hamed, *Cascaded Interferometers and Their Medical Applications*,
SpringerBriefs in Applied Sciences and Technology,
https://doi.org/10.1007/978-3-031-64535-8_1

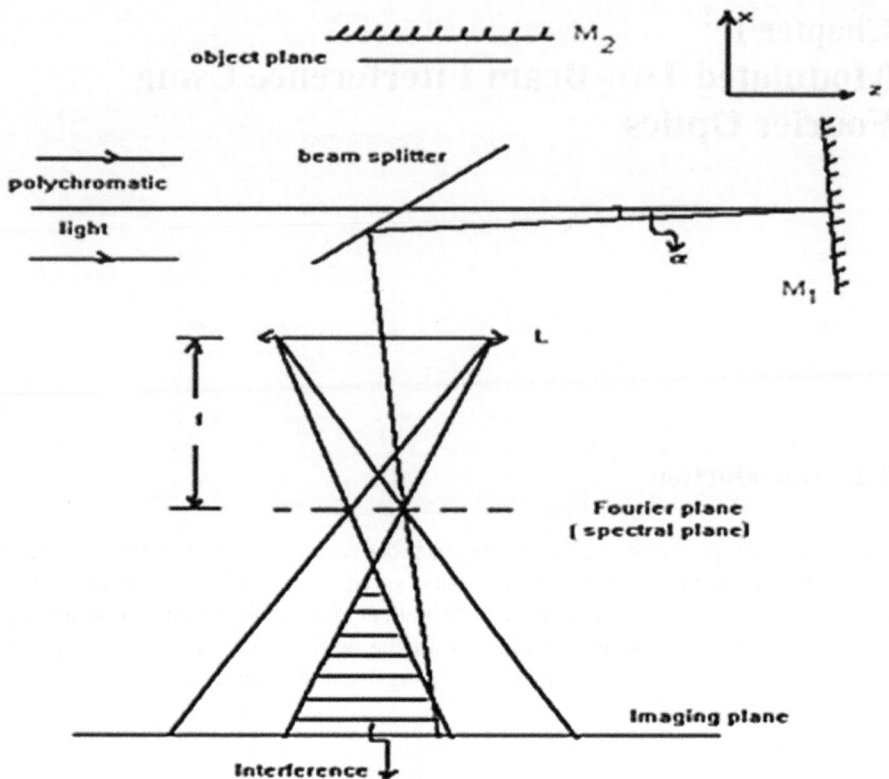

Fig. 1.1 Two-beam interference using polychromatic light where a phase object is introduced in arm M_2. α is the oblique angle between the incident ray on mirror M_1 and the reflected ray, and L is the imaging lens

$$A_0(x, y, \lambda) = \beta_2 a(\lambda) \exp\left[-j\Phi_0(\lambda)\right]; \quad \beta_2 = r_s.t_s.r_2 \tag{1.2}$$

r_2 is the amplitude reflection coefficient of mirror M_2.

In the Fourier plane, as shown in Fig. 1.1, the inclined plane wave represented by Eq. (1.1) is transformed to a shifted Dirac delta function located at the point $v = -f \sin(\alpha)$ in the plane (u, v) as follows:

$$\text{F.T.}\{A_c(x, y)\} = \text{F.T.}\left\{\beta_1 a(\lambda) \exp\left[-j\left(\frac{2\pi}{\lambda}\right)y\sin(\alpha)\right]\right\}$$

$$= \beta_1 a(\lambda) \int\limits_{-\infty}^{\infty}\!\!\int \exp\left[-j\left(\frac{2\pi}{\lambda}\right)y\sin(\alpha)\right]\exp\left[-j\left(\frac{2\pi}{\lambda f}\right)(xu + yv)\right]\mathrm{d}x\,\mathrm{d}y$$

$$= \beta_1 a(\lambda) \int\limits_{-\infty}^{\infty} \int \exp\left\{-j\left(\frac{2\pi}{\lambda f}\right)[xu + y(v + f\ \sin\alpha)]\right\}dx\ dy$$

$$= \beta_1 a(\lambda).\delta(u, v + f\ \sin\alpha) \tag{1.3}$$

Similarly, the Fourier transform of the object, using monochromatic light for illumination, is represented as follows:

$$\tilde{A}_0(u, v, \lambda) = \text{F.T.}\{A_0(x, y, \lambda)\} \tag{1.4}$$

This can be represented as the Fourier spectrum of the object convoluted with a Dirac delta function located at the point $(0, 0)$ in the Fourier plane (u, v). It is rewritten as follows:

$$\tilde{A}_0(u, v, \lambda) = \tilde{A}_0(u, v, \lambda) \otimes \delta(u, v) \tag{1.5}$$

The addition of Eqs. (1.3) and (1.5) in the imaging plane (x', y') gives the interference pattern originating from the inverse Fourier transform as follows:

The optical phase difference is calculated as follows:

$$\phi_0(\lambda) = \frac{2\pi}{\lambda}(O.P.D.) \tag{1.6}$$

$O.P.D. = 2\,[n(\lambda) - 1]\,t_p$ is the optical path difference, and t_p is the thickness of the object.

In the imaging plane of the coordinates (x', y'), the intensity distribution is calculated by taking the modulus square of the transmitted complex amplitudes of both the object transparency and the carrier wave, i.e.,

$$I(x, y) = |Ac(x, y) + Ao(x, y)|^2 = |A_c|^2 + |A_0|^2 + A_c A_0^* + A_c^* A_0 \tag{1.7}$$

$(*)$ is used as a symbol for complex numbers.

By substituting Eqs. (1.1) and (1.2) into (1.7), we finally obtain the following result:

$$I(x, y) = |a(\lambda)|^2 \left\{T_s R_s R_1 + T_s R_s R_2 + 2T_s R_s\,(R_1 R_2)^{\frac{1}{2}}\ cos\left[\phi_0(\lambda) - \left(\frac{2\pi}{\lambda}\right)y\ \sin\alpha\right]\right\} \tag{1.8}$$

where $R_i = |r_i|^2$, $i = 1, 2$ and $T_s = |t_s|^2$.

If R_1 and R_2 have equal reflectivity, we obtain the well-known expression of the two-beam interference as:

$$I(x, y) = 4\gamma \beta_1\ cos^2\left[\frac{\phi_0}{2} - \left(\frac{\pi}{\lambda}\right)y\ \sin\alpha\right] \tag{1.9}$$

$\gamma = |a(\lambda)|^2$ is characteristic of the illumination of the interferometer, and $\beta_1 = \beta_2 = T_s R_s R_1$ is dependent upon the optical components of the interferometer, namely the beam splitter and the mirrors M_1 and M_2.

1.2.1 Rectangular Model for the Fringe Shift

This means the utilization of an object of a constant thickness t_p and having a uniform refractive index (n). This can be represented by a rectangular function as follows:

$$n(y) = \text{rect}(y) = n(\lambda); y \le a$$
$$= 1; y > a \tag{1.10}$$

In this case, of monochromatic illumination and by substitution of Eqs. (1.10) in (1.9), we obtain the following intensity distribution:

$$I(x, y) = a^2(\lambda) T_s R_s \, R_1 \, \cos^2\left\{\left(\frac{\pi}{\lambda}\right)[y \sin\alpha - 2(n - 1)t_p]\right\} \tag{1.11}$$

The phase of the object is calculated for the rectangular object as:

$$\phi_0(\lambda) = \left(\frac{2\pi}{\lambda}\right)[2(n - 1)t_p] \tag{1.12}$$

To compute the fringe shift introduced by the phase object, we first calculate the fringe spacing ΔZ from the difference between two consecutive maxima. Using Eq. (1.11), we obtain:

$$\Delta Z = \frac{\lambda}{2 \sin\left(\frac{\alpha}{2}\right)} \tag{1.13}$$

If the shift introduced by the phase object is less than the interfringe spacing, i.e., $\Delta S < \Delta Z$, then we can write this inequality:

$$t_p < \frac{\lambda}{4(n - 1) \sin\left(\frac{\alpha}{2}\right)} \tag{1.14}$$

$n = n(\lambda)$, is constant for monochromatic illumination.

Hence, the differential fringe shift is calculated as follows:

$$D_x(y) = \frac{\Delta S}{\Delta Z} = 2(n - 1)\left(\frac{t_p}{\lambda}\right) \sin\left(\frac{\alpha}{2}\right) \tag{1.15}$$

This differential fringe shift may be introduced in the Abel inversion formula to map the phase distribution of the object.

1.2.2 Triangular Model for the Fringe Shift

This means the utilization of an inhomogeneous phase object of constant thickness (t_p) but with a variable refractive index. The variation is represented by a triangular function represented as

$$n(y) = n_0\left[1 - \left|\frac{(y - y_0)}{b}\right|\right]; \ |y| \le b$$
$$= 1; \ |y| > b \qquad (1.16)$$

n_0 is the maximum refractive index located at the center of the pattern, and y_0 represents the shift introduced at a certain depth x. Equation (1.16) can be rewritten symbolically as a convolution product of a triangular function located at the center and a Dirac delta function representing the shift.

$$n(y) = n_0\left[1 - \left|\frac{y}{b}\right|\right] \oplus \delta(y - y_0) \qquad (1.17)$$

where $\otimes$ is a symbol for the convolution product and δ is the Dirac delta function.
From Eqs. (1.16), (1.9), and (1.6), we obtain:

$$I(x, y) = 4a^2(\lambda)T_s R_s R_1 \cos^2\left\{\left(\frac{\pi}{\lambda}\right)\left[y \sin\alpha - 2(n(y) - 1)t_p\right]\right\} \qquad (1.18)$$

Using Eqs. (1.14) and (1.15), the fringe shift may be calculated from the equation of straight-line fringes modulated by a triangular fringe shift as follows:

$$m\lambda = y \sin\alpha - 2\left(n_0(1 - \left|\frac{y}{a}\right|)t_p - 2n_0 y_0\frac{t_p}{b} + 2t_p \qquad (1.19)$$

If this condition is fulfilled: $n_0 y_0 = b$. For example, if $n_0 = 1.5$ and $b/y_0 = 1.5$. In this case, Eq. (1.19) becomes:

$$m\lambda = y \sin\alpha - 2\left(n_0(1 - \left|\frac{y}{a}\right|)t_p \qquad (1.20)$$

Consequently, the differential fringe shift becomes:

$$D_x(y) = \frac{\Delta S}{\Delta Z} = 4n_0\left(1 - \left|\frac{y}{a}\right|\right)\left(\frac{t_p}{\lambda}\right)\sin\left(\frac{\alpha}{2}\right) \qquad (1.21)$$

It represents an exact triangular function for the cited model, and Δz is the inter fringe spacing calculated from Eq. (1.13).

1.2.3   A Truncated Gaussian Model for the Fringe Shift

A truncated Gaussian function is assumed to represent the fringe shift obtained in the case of two-beam interference. An inhomogeneous phase object of a refractive index $n(y)$ is assumed to have a shifted Gaussian function, i.e.,

$$n(y; \lambda) = n_0 \exp\left\{-\beta\left[\frac{y - y_0}{w}\right]^2\right\}; \ |y| \leq w \qquad (1.22)$$

where w is the truncation width, β is a parameter, and y_0 represents the shift introduced at a certain depth x. As before, Eq. (1.22) can be represented by the following convolution product:

$$n(y; \lambda) = n_0 \exp\left\{-\beta\left[\frac{y}{w}\right]^2\right\} \otimes \delta(y - y_0) \qquad (1.23)$$

In this case, the intensity distribution is Eq. (1.18), except that the refractive index is represented by Eq. (1.22). The straight-line fringes are modulated by Gaussian distribution, which represents the proposed shift represented as:

$$m\lambda = y \sin \alpha - 2\{n_0 \exp\left\{-\beta\left[\frac{y - y_0}{w}\right]^2\right\} - 1\} t_p \qquad (1.24)$$

The shift y_0 may be compensated for by the linear shift (t_p), leading to an exact Gaussian distribution for the fringe shift.

Hence, the differential fringe shift is calculated as follows:

$$D_x(y) = \frac{\Delta S}{\Delta z} = 2\left(\frac{t_p}{\lambda}\right) n_0 \exp\left\{-\beta\left[\frac{y - y_0}{w}\right]^2\right\} \sin \alpha/2 \qquad (1.25)$$

1.3   Results and Discussion

A computer program is constructed to map the straight-line fringes obtained from the two-beam interference modulated by the fringe shift models. Figure 1.2 shows the relationship obtained between the order of interference (m) and the coordinate

(y), considering a uniform refractive index of $n = 1.5$ represented by the shifted rectangular function from the straight-line fringes located symmetrically around the z-axis. In this case, the fringe shift is constant. Forty straight-line fringes are plotted in this Figure 1.2. The corresponding refractive indices are plotted in Figure 1.3.

The second model of the triangular function is plotted in Fig. 1.4, where thirty-five fringes are plotted, and the modulation of the triangular function is centered symmetrically around the z-axis. A magnified portion of only eight fringes is plotted in Fig. 1.5, and the corresponding refractive index is shown in Fig. 1.6.

The third model of the truncated Gaussian function is plotted in Fig. 1.7, and the corresponding refractive index distribution is plotted in Fig. 1.8. These results show that the distribution of the refractive indices is like that of the above-mentioned

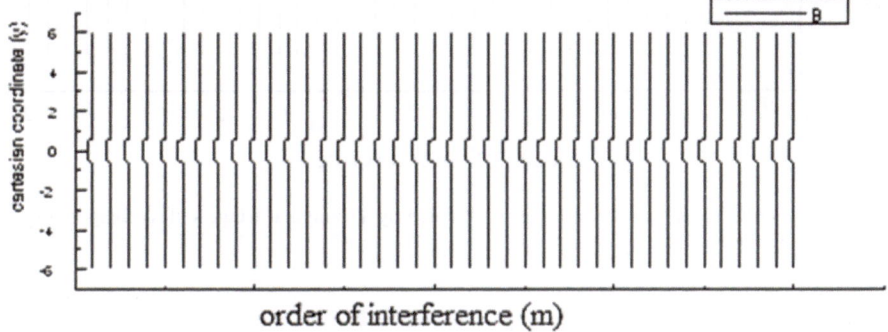

Fig. 1.2 Straight-line interference fringes modulated by a rectangular fringe shift corresponding to the 1st rectangular model. Forty fringes are shown

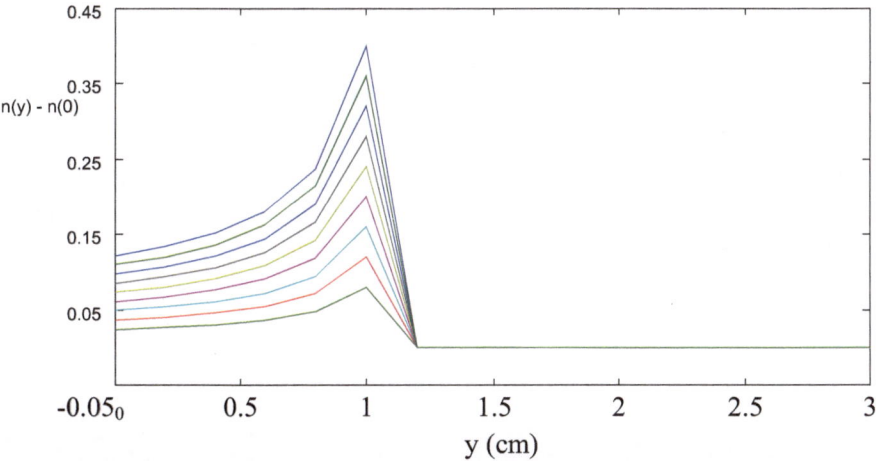

Fig. 1.3 Theoretical contour mapping of the refractive index using a rectangular window function for the input data

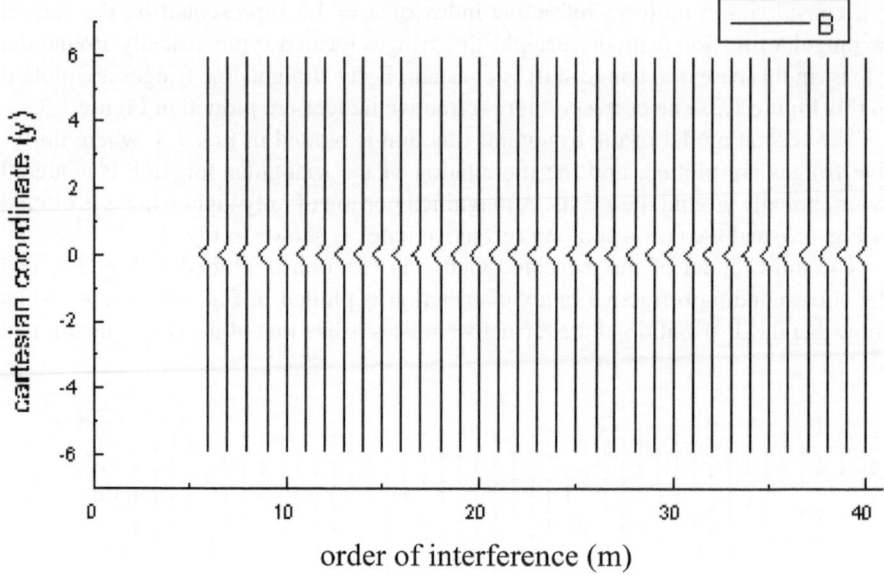

Fig. 1.4 Straight-line interference fringes modulated by a triangular fringe shift corresponding to the 2nd triangular model

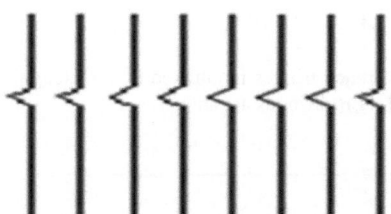

Fig. 1.5 A modified portion of eight fringes modulated by a triangular fringe shift (Δs)

models, may be attributed to the manipulation of imaging interference. The theoretical curves of the refractive indices are computed from the back-substitution simulation process.

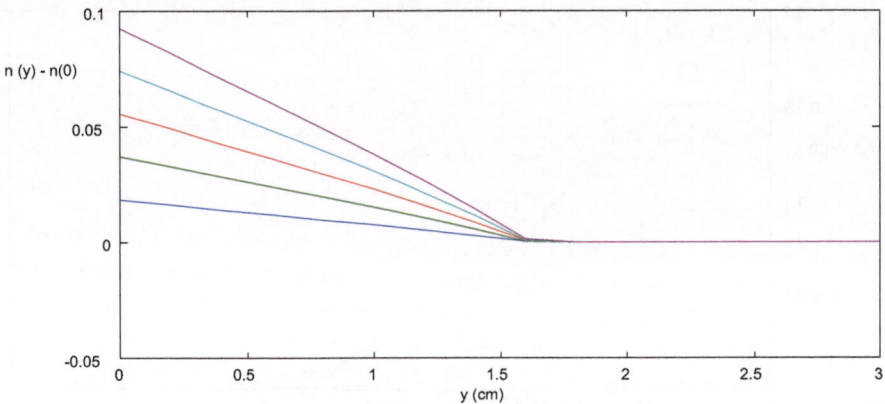

Fig. 1.6 Theoretical contour mapping of the refractive index versus the coordinate (y) using a triangular input function

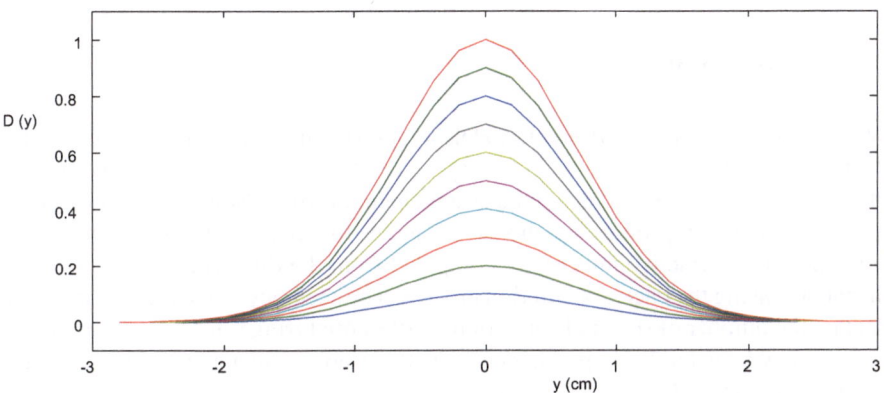

Fig. 1.7 Truncated Gauss function used as input data for the processing of the refractive index distribution of a thermal source

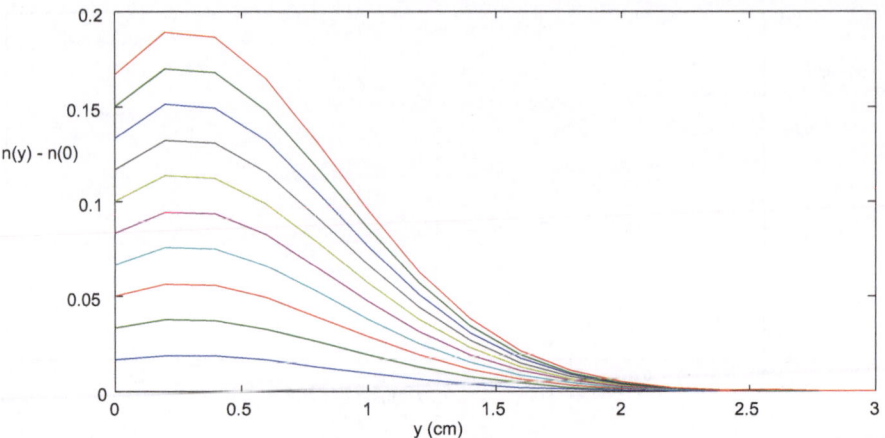

Fig. 1.8 Theoretical contour mapping of the refractive index using a truncated Gauss function for the input data

1.4   Conclusion

We have suggested three different models to describe the fringe shifts introduced in the phase term that appear in the intensity distribution. In this study, the interference of two beams is considered using the arrangement of Michelson. Rectangular, triangular, and Gaussian profiles are chosen as models to represent the fringe shifts. The corresponding refractive indices are computed from the differential fringe shifts of the models using the back-substitution process. The accuracy of the method is dependent on the number of selected zones of the differential fringe shifts. This simulation process may be extended to process colored phase objects using polychromatic light for illumination of the interferometer.

References

1. A.M. Hamed, Computer simulation of modulated two-beam interference using monochromatic light. Opt. Applicata **34**, 51–61 (2004)
2. N. Barakat, S. Mokhtar, J.O.S.A. **53**, 159 (1963)
3. M. Born, E. Wolf, *Principles of Optics* (Macmillan Co., New York, 1968)
4. R.N. Bracewell, *The Fourier Transform and its Applications* (McGraw-Hill Co. New York, 1978)
5. S.G. Lipson, H. Lipson, D.S. Tannhauser, *Optical Physics* (Cambridge Univ. Press, 1995)
6. J.W. Goodman, *Introduction to Fourier Optics*, 2nd ed. (McGraw-Hill, 1996)
7. A.M. Hamed, Master of Science (M.Sc.) Thesis, A. R. E. (1976), Some applications of scattered light and multiple beam interference using coherent light
8. A.M. Hamed, Fourier imaging of uncladded fibers using a liquid wedge interferometer. Opt. Applicata **27**, 229–240 (1997)
9. A.M. Hamed, *Fourier Optics, Laser Speckle Imaging, and Cascaded Interferometers* (Lambert Academic Publishing (LAP), 29 September 2020)

Chapter 2
A Cascaded Michelson Interferometer and Its Application to Glass Fibers

We propose cascaded two-beam interference, which is considered a cosine function of higher order n greater than one ($n > 1$). We investigate the fringe shift of the microscopic object w.r.t the cascaded two-beam interference of higher orders of $n = 1, 2, 3$ up to $n = 25$. The fringe sharpness of the cascaded higher-order two-beam interference is compared with that of ordinary two-beam interference. Modulated multiple beam interference is compared with the improved two-beam model for $n = 25$. The refractive index of the microscopic optical fiber is extracted from the fringe shift of the improved cascaded two-beam interference. MATLAB code was used to fabricate the investigated interference-modulated images.

2.1 Introduction

Two-beam interference originates from either the division of amplitude or the wave-front giving an intensity distribution proportional to $\cos^2\theta$, as in Michelson or other two-beam interferometers. The sharpness is improved using multiple beam inter-ference, as in the Fabry–Perot interferometer, given the well-known Airy distri-bution formula. Many applications based on measuring the fringe shift in optical and synthetic fibers to obtain refractive index information [1–11, 14] are outlined. Recently, digital two- and multiple beam interference has been applied to coronavirus images and other medical images [12, 13] to extract the refractive index, which is related to fringe shifts.

Now, in the present chapter, we suggest a cascaded two-beam interference. The intensity is in the form of a higher-order function in the form $\cos^{2n}(\theta)$, where n repre-sents the number of feedback passes to the interferometer. This sharpens the fringes to sharper fringes than the corresponding sharpness in the case of ordinary two-beam interference. A cascaded two-beam interferometer is presented, followed by a theoretical analysis. The results and discussion are given, followed by a conclusion.

© The Author(s), under exclusive license to Springer Nature Switzerland AG 2024   13
A. Hamed, *Cascaded Interferometers and Their Medical Applications*,
SpringerBriefs in Applied Sciences and Technology,
https://doi.org/10.1007/978-3-031-64535-8_2

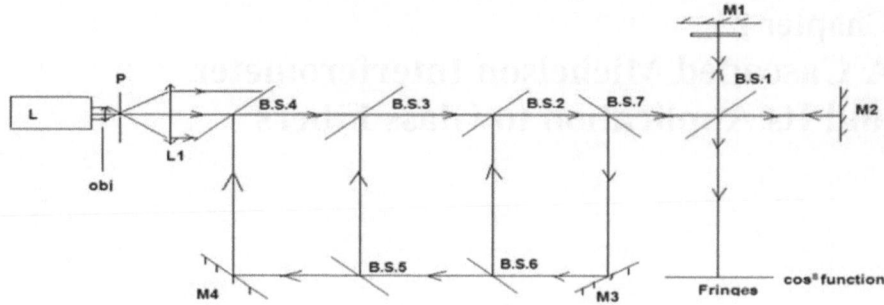

Higher Order two beam interference arrangement of multiples of cos² function.

Fig. 2.1 A cascaded higher-order two-beam interference where three loops are added to the original arrangement, giving the $\cos^8$ function

2.2   The Cascaded Two-Beam Interference is Based on the Division of the Amplitude

As shown in Fig. 2.1, L is a laser beam with a monochromatic wavelength of $\lambda = 633$ nm. Obj.: low numerical objective lens of $NA = 0.5$. P: The pinhole placed in the focal plane of the objective lens of diameter allows only the central band of the diffraction pattern, and L_1 is placed at a distance from the pinhole = focal length of L_1. Parallel rays of uniform intensity originate from the Gaussian beam. M_1 and M_2 are reflecting mirrors corresponding to the original Michelson interferometer, where the object is placed in the path of the beam reflected by mirror M_1. In the absence of these feedback loops, only the $\cos^2$ function is captured by the detector in the observation plane.

Now, the presence of feedback mirrors M_3 and M_4, and beam splitters B.S.$_2$—B.S.$_6$ allow an attack on the interferometer three times. Hence, three loops corresponding to the three feedback passes allow incidence on the interferometer, giving $(\cos^2)^n = \cos^6$ function in addition to the original $\cos^2$ function of the interferometer. Consequently, the $\cos^8$ functional group is fabricated in the observation imaging plane. Hence, more feedback passes of the beam further improve the sharpness compared with the ordinary two-beam $\cos^2$ function.

2.3   Theoretical Analysis

The coherent multiplication of two-beam interference is governed by the feedback of the coherent laser beam incident on the Michelson arrangement. Hence, the number of feedback passes (N) on the interferometer gives an intensity distribution in the form:

$$I_{\text{feed back}}(x, y; N) = I_0\cos^{2(N+1)}(\delta)$$ (2.1)

The ordinary two-beam interference has an intensity distribution extracted from Eq. (2.1), where $N = 0$ (no feedback), as follows:

$$I_{\text{ordinary}}(x, y; 0) = I_0\cos^2(\delta)$$ (2.2)

For a single feedback, $N = 1$

$$I_{\text{feed back}}(x, y; N = 1) = I_0\cos^4(\delta)$$ (2.3)

For two feedback light passes, $N = 2$

$$I_{\text{feed back}}(x, y; N = 2) = I_0\cos^6(\delta)$$ (2.4)

It is known that each feedback pass adds $\cos^2\delta$ to the original $\cos^2\delta$.

The fringe sharpness is conveniently measured by the full intensity width at half maximum (FWHM) [15]. In the case of ordinary two-beam interference, the fringe sharpness is computed as:

$$\frac{I}{I_0} = \frac{1}{2} = \cos^2(\delta_w)$$ (2.5)

In this case, the FWHM represented by δ is computed as follows:

$$\delta_w = \cos^{-1}\left(\frac{1}{\sqrt{2}}\right)$$ (2.6)

In the case of multiple feedback of the number of passes N, the FWHM is computed as follows:

$$\delta_w = \cos^{-1}\left\{\frac{1}{\frac{1}{(N+1)}[\sqrt{2}]}\right\}$$ (2.7)

For $N = 0$, no feedback, Eq. (2.7) reduces to Eq. (2.6).

For $N = 1$,

$$\delta_w = \cos^{-1}\left\{\frac{1}{\frac{1}{2}[\sqrt{2}]}\right\}$$

For $N = 2$,

$$\delta_w = \cos^{-1}\left\{\frac{1}{\frac{1}{3}[\sqrt{2}]}\right\}$$

etc.

If we consider multiple reflections of the reference beam over the object beam, multiple beam interference is modulated by the object information. In this case, the intensity distribution of multiple beam interference is well known as the Airy pattern and is represented as follows:

$$I(x, y, z) = I_0 \frac{1}{1 + F \sin^2(\frac{\delta}{2})}; \delta = \left(\frac{2\pi}{\lambda}\right) O.P.D. \tag{2.8}$$

where δ is the phase difference between the object beam and multiple reference beams and O.P.D. is the corresponding optical path difference.

Equation (2.8) is written in discrete form as follows:

$$I(x, y, z) = I_0 \sum_{n=1}^{N} \sum_{m=1}^{M} \left\{ \frac{1}{1 + F \sin^2\left(\frac{\Phi(n\Delta x, m\Delta y; z) - \Psi(n\Delta x, m\Delta y)}{2}\right)} \right\} \tag{2.9}$$

The fringe width in the case of multiple beam interference is given in [16] as follows:

$$\delta_w = \frac{4}{\sqrt{F}} \quad ; F = 4R/(1 - R)^2 \tag{2.10}$$

R is the reflectivity of the plate surface. The fringe sharpness is improved for higher reflectivity.

2.4 Computation of the Refractive Indices of Microscopy Images Using the Modified Michelson Interferometer

The refractive index of any microscopy image μ in the case of a modified Michelson interferometer is computed as follows:

Since the phase of the wave cumulates traveling a distance L in a medium is

$$\delta(x, y) = \int_L k \, dl = \frac{2\pi}{\lambda} \int \mu(x, y, z) dl \tag{2.11}$$

Then, the same wave that propagates over two equivalent paths L in the microscopic object and the vacuum gives the phase difference as follows (Fig. 2.2):

$$\Delta\delta(x, y; z) = 2\pi/\lambda \int [\mu(x, y, z) - 1] dl \tag{2.12}$$

Fig. 2.2 Propagation of light in the object of refractive index μ compared with air

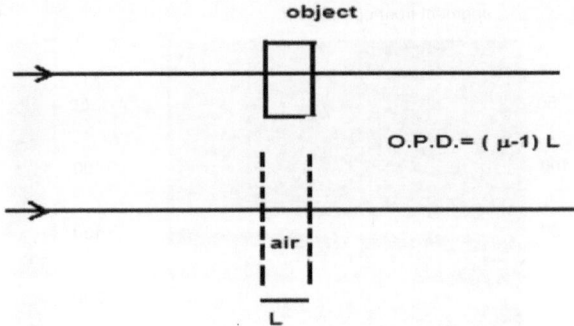

where $k = \omega/c = 2\pi/\lambda$ is the propagation wavenumber in a medium with refractive index μ and k_0 is the propagation constant in vacuum.

Finally, from Eq. (9.12), the refractive index of the microscopy image is obtained as follows [13, 16, 17]:

$$\mu(x, y)_{const\,x} = 1 + a(x, y)\frac{\delta z}{\Delta z} \tag{2.13}$$

The fringe shift is δz concerning the inter fringe spacing Δz at constant x, the fringes are assumed to be in the x-y plane, z is the axis normal to the fringe system which represents the height depth, and a (x, y) represents the amplitude of the image. In Eq. (2.13),

$$h(x, y, z) = a(x, y).\delta z. \tag{2.14}$$

2.5 Results and Discussion

A segment from the optical fiber and the corresponding modulated two-beam interference are shown in Fig. 2.3. Both images have dimensions of 256×256 pixels. The modulated two beam feedback interference is shown in Fig. 2.4. The image on the right has a $\cos^{10}\delta$ originating from 4 feedback passes ($N = 4$) compared with the ordinary $\cos^2\delta$. It is shown by the naked eye that the multiple feedback passes on two-beam interference improve the sharpness of the fringes compared with the ordinary two-beam interference. A further improvement in the fringe sharpness is attained by increasing the number of feed-back passes, as shown in Fig. 2.5, where $N = 9$ and $N = 24$. In addition, the two beam fringes with a high feedback pass of $N = 24$ are compared with the ordinary multiple beam interference, as shown in Fig. 2.6. The fringe sharpness in the case of multiple feedback passes in two beam arrangements is computed using Eq. (2.7), while the ordinary fringe sharpness is computed from Eq. (2.6).

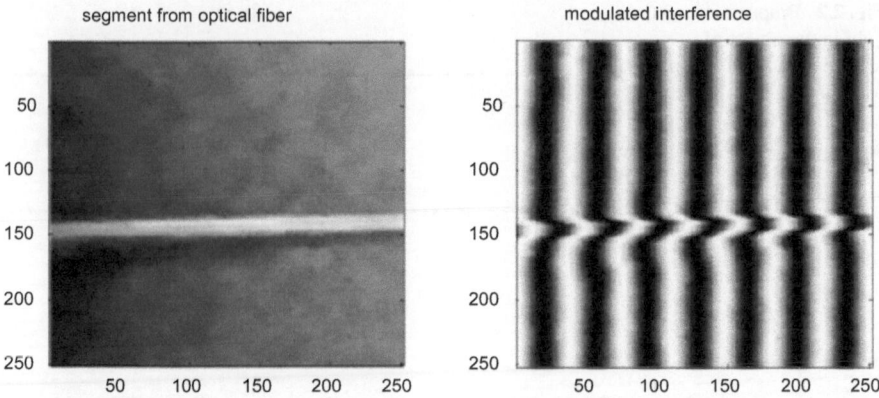

Fig. 2.3 A segment from an optical fiber and the corresponding modulated two-beam interference. Both images have dimensions of 256 × 256 pixels

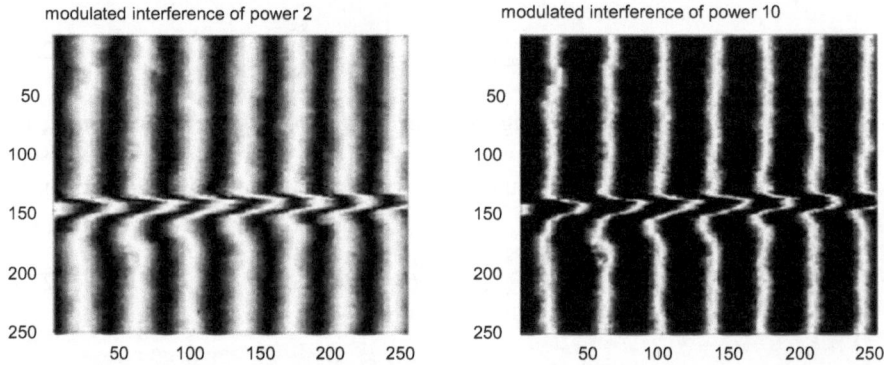

Fig. 2.4 Modulated two-beam interference. The image on the left represents the ordinary $\cos^2\delta$ $N = 0$, while the image on the right has $\cos^{10}\delta$ where $N = 4$

Some results of the fringe sharpness computed from the FWHM using Eq. (2.7) are constructed as in Table 2.1. The sharpness in the case of feedback passes in the modified Michelson arrangement is much sharper than the corresponding sharpness in ordinary two-beam interference. This sharp improvement is dependent on the number of feedback passes N. For example, the sharpness is reduced from 45 for an ordinary two-beam beam to 14.9978 for multiple feedback passes, where $N = 9$ for $I = I_0 \cos^{2(N + 1)} \delta = I_0 \cos^{20} \delta$.

The refractive index of the fiber is computed from equation (2.13), where a $(x, y) = 1$, and the average interfringe spacing is computed from the image shown in Fig. 2.7 to give $\Delta Z = 78$ pixels. The fringe shift $\delta Z = Z - Z_{\text{image}}$ is computed for six fringes taken from the same Fig. 2.7. Finally, the average refractive index taken from Table 2.2 is computed to obtain:

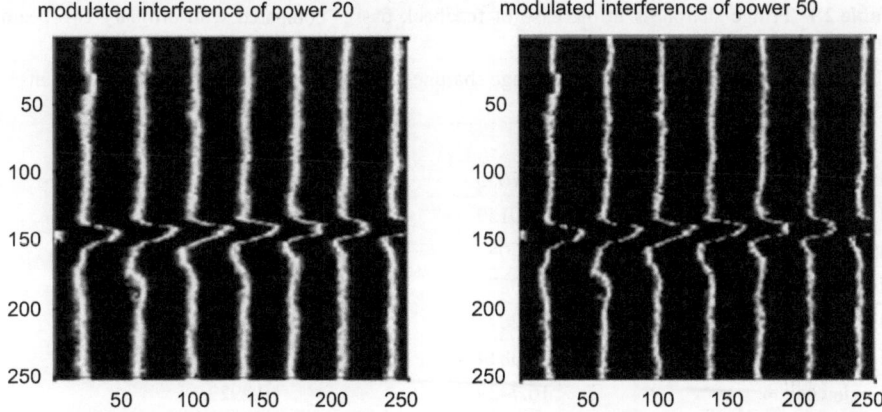

Fig. 2.5 Modulated two-beam interference. The image on the left represents $\cos^{20}\delta$ $N = 9$, while the image on the right represents $\cos^{50}\delta$ where $N = 24$

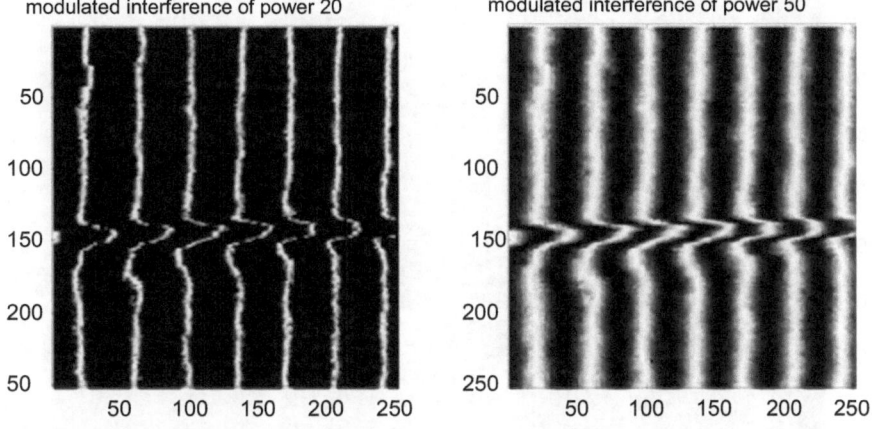

Fig. 2.6 Modulated interference. The image on the left represents two-beam $\cos^{50}\delta$ interference for $N = 24$, while the image on the right has ordinary multiple beam interference. In all modified two-beam images, the intensity is in the following form: $I = I_0 \cos^{2(N+1)}(\delta)$

$$< \mu >= (1/M) \sum_{m=1}^{M} \mu(Z) = 1.5976.$$

Figure 2.8 shows the fringe shift of the fiber in multiple beam interference, where 14 fringes are shown in the image of dimensions (512×512) pixels using the Airy distribution formula represented by Eq. (2.9).

Another application of metal inspection using multiple beam interference is shown. An image of water droplets agglomerated on an aluminum surface with

Table 2.1 Fringe sharpness in the case of feedback passes compared with ordinary two-beam interference in a Michelson arrangement

The two-beam intensity with feedback	N	Fringe sharpness in degrees	Fringe sharpness in radians
$I = I_0 \cos^2 \delta$	0	45	0.7853
$I = I_0 \cos^4 \delta$	1	32.7655	0.57186
$I = I_0 \cos^6 \delta$	2	27.0139	0.4715
$I = I_0 \cos^8 \delta$	3	23.5083	0.4103
$I = I_0 \cos^{10} \delta$	4	21.087	0.3680
$I = I_0 \cos^{12} \delta$	5	19.2875	0.3366
$I = I_0 \cos^{14} \delta$	6	17.8814	0.3121
$I = I_0 \cos^{16} \delta$	7	16.7439	0.2922
$I = I_0 \cos^{18} \delta$	8	15.7990	0.2757
$I = I_0 \cos^{20} \delta$	9	14.9979	0.2617

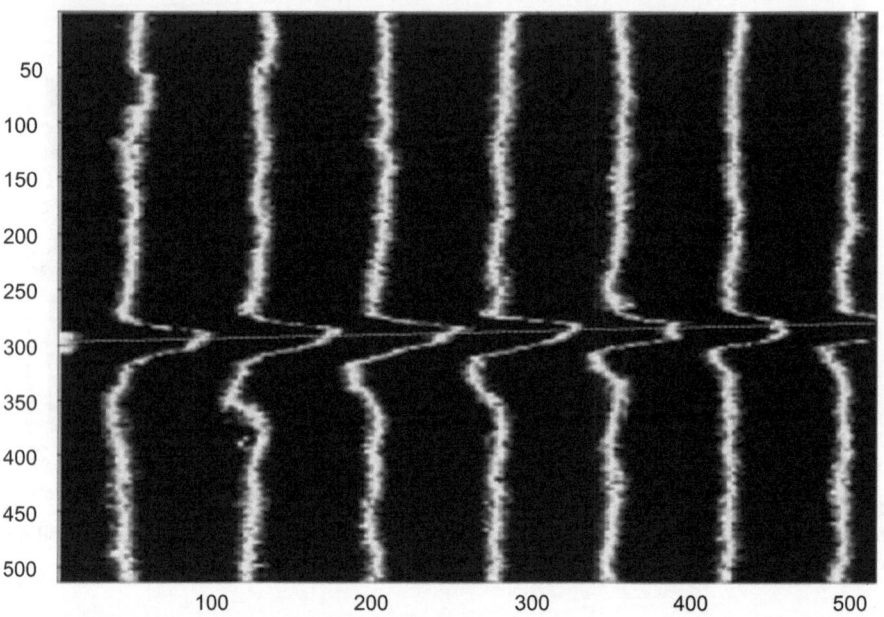

Fig. 2.7 Image represents two beams in the form $I = I_0 \cos^{50}\delta$ for $N = 24$ used in the computation of the refractive index of the fiber segment. The discontinuous line is adjusted over the fringe shift to obtain accurate values. The image has matrix dimensions of 512×512 pixels. The computations applied on only six fringes are shown in the image

Table 2.2 Refractive index values as a function of the Z coordinate at a certain line adjusted over the fringe shift at (297,280) pixels

Z	Z image	$\delta Z = Z—Z$ image	$\mu(Z) = 1 + \delta Z/\Delta Z$
40	90	50	1.6757
117	172	55	1.7051
197	248	51	1.6538
274	320	46	1.5897
344	384	40	1.5128
420	455	35	1.4487

$\Delta Z = 78$ pixels represents the average interfringe spacing. The shift of the modulated fringes is toward the right

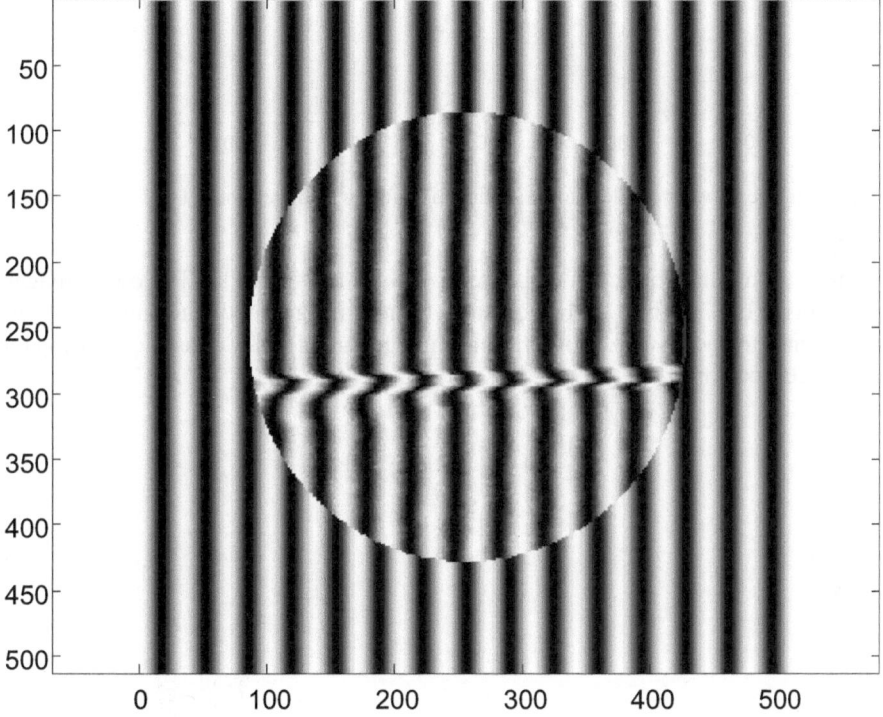

Fig. 2.8 Fringe shift of the fiber in multiple beam interference, where 14 fringes are shown in the image with 512×512 pixels

dimensions of 512×512 pixels is shown in Fig. 2.9. The corresponding multiple beam interference is plotted in Fig. 2.10, where the random irregular fringe shifts corresponding to the agglomerated droplets are shown. In the upper left, a segment from the whole image at $(i,j) = (50–150\,\text{pixels}, j = 200–300\,\text{pixels})$ where the droplet

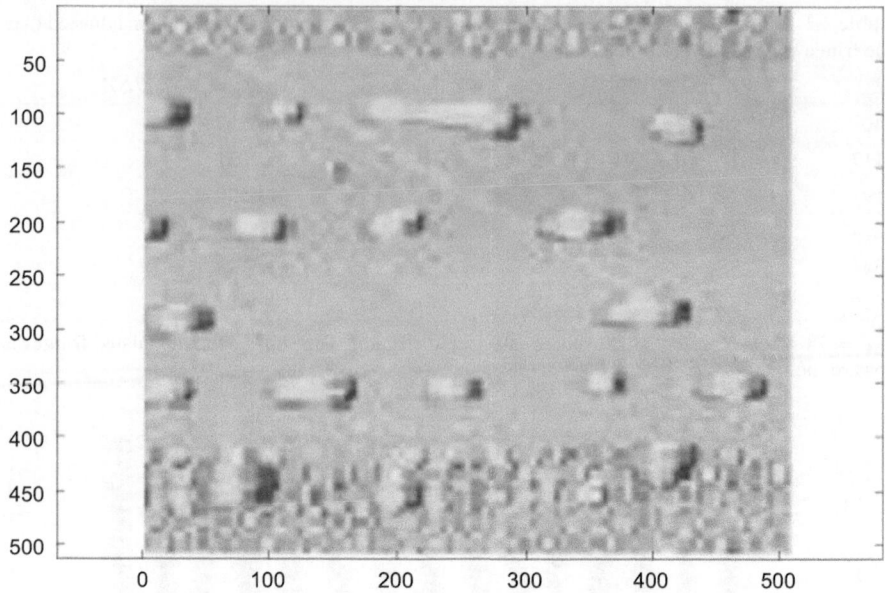

Fig. 2.9 Image of water droplets on an aluminum surface with dimensions of 512×512 pixels

from the aluminum surface is magnified is shown in Fig. 2.11a. The other images, shown in the Fig. 2.11b–d, correspond to 3, 6, and 9 fringes modulated multiple beam interference, respectively. All interference images are digitally constructed using the MATLAB code used in [13].

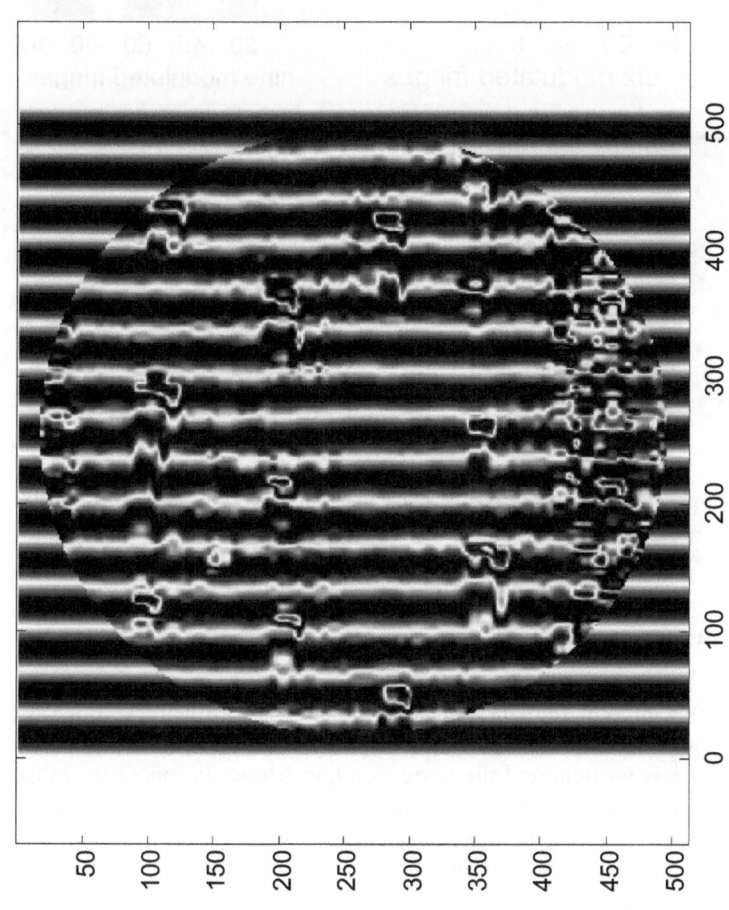

Fig. 2.10 Modulated multiple beam interference fringes of water droplets on the aluminum surface using high reflectivity coating mirrors with $R = 0.8$

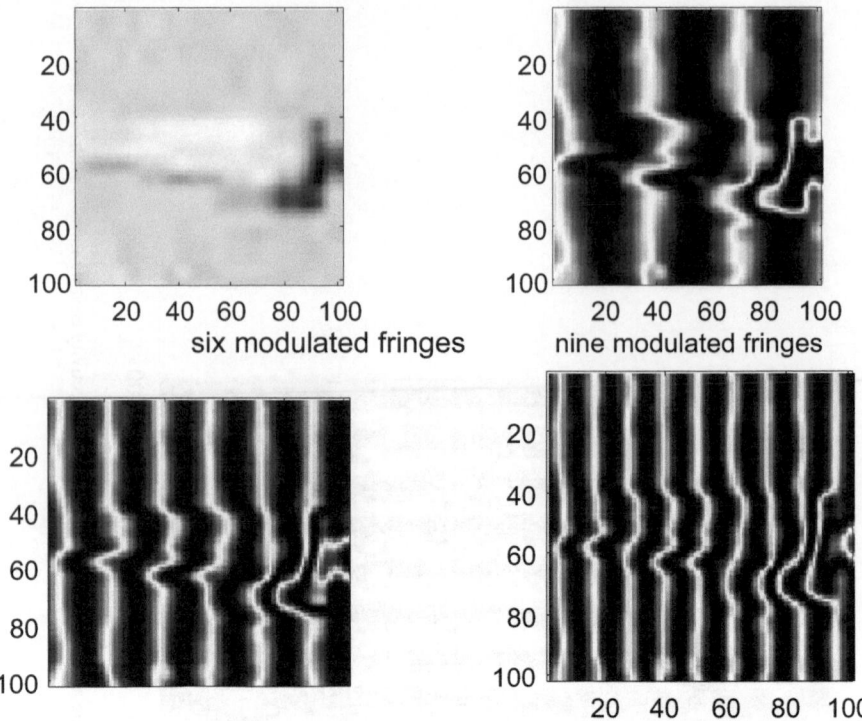

Fig. 2.11 In the upper left, a segment from the whole image at $(i, j) = (50\text{--}150 \text{ pixels}, j = 200\text{--}300$ pixels) where a magnified droplet from the aluminum surface is shown. The other images correspond to 3, 6, and 9 fringes of modulated multiple beam interference

2.6 Conclusion

The feedback of coherent light is suggested in the two-beam interference of the Michelson arrangement. The intensity distribution in the case of feedback of two-beam interference is written, and the corresponding fringe sharpness is computed. A comparison with ordinary two-beam fringe sharpness is discussed. The fringe sharpness in the case of feedback light passes is better than the corresponding fringe sharpness in the case of two-beam interference. In addition, the fringe sharpness further improved with increasing number of feedback passes. Finally, compared with that of ordinary two-beam interference, the refractive indices of microscopy images are more accurate. Consequently, the final goal of this chapter based on three objectives is realized. The first is the improvement in fringe sharpness in the case of the proposed modified Michelson interferometer. The 2nd is the accurate measurement of the fringe shift computed from the considered sharp fringes. The 3rd is the extraction of the refractive index of the fiber specimen from the fringe shift results.

References

Journal References

1. N. Barakat, S. Mokhtar, On the factors contributing to the formation of multiple beam Fizeau fringes. J. Opt. Soc. Am. **53**, 300–301 (1963)
2. N. Barakat, A.M. Hindelah, Determination of the refractive indices, birefringence, and tensile properties of normal viscose rayon fibers. Textile Res. J. **41**, 581–584 (1964)
3. N. Barakat, Interferometric studies on fibers: part I: theory of interferometric determination of indices of fibers. Textile Res. J. **41**, 167 (1971)
4. M.M. El Nicklawy, I.M. Fouda, An analysis of Fizeau fringes crossing a fiber with multiple skins. Textile Res. J. **71**, 252–256 (1980)
5. A. Hamza, T.Z.N. Sokkar, M.A. Abeel, Interferometric determination of optical properties of fibers of irregular transverse section and having a skin- core structure. J. Phys. D **18**, 1773 (1985)
6. A. Hamza, T.Z.N. Sokkar, M.A. Abeel, Interferometric determination of refractive indices and birefringence of fibers with irregular transverse sections. J. Phys. D **19**, 119 (1986)
7. N. Barakat, A. Hamza, A. Goned, Multiple beam interference fringes applied to GRIN optical waveguides to determine fiber characteristics. Appl. Opt. **24**, 4383–4386 (1985)
8. L.M. Boggs, H.M. Presby, D. Marcuse, Rapid automatic index profiling of whole fiber samples: part I. Bell System Tech. J. **58**, 867–882 (1979)
9. H.M. Presby, D. Marcuse, H.W. Astle, L.M. Boggs, Rapid automatic index profiling of whole fiber samples: part II. Bell System Tech. J. **58**, 883–902 (1979)
10. A.M. Hamed, Fourier imaging of uncladded fibers using a liquid wedge interferometer. Opt. Appl. **27**, 229–240 (1997)
11. A.M. Hamed, Modeling of the fringe shift in multiple beam interference for glass fibers. Pramana J. Phys. **70**, 643–648 (2008)
12. A.M Hamed, Investigation of SIDA virus (HIV) images using interferometry and speckle techniques. Int. J. Innovative Res. Comp. Sci. Tech. (IJIRCST) **4**, 38–45 (2016)
13. A.M. Hamed, Image processing of coronavirus using interferometry. Optics Photon. J. **6**, 75–86 (2016)

Book References

14. N. Barakat, A. Hamza, *Interferometry of Fibrous Materials (Bristol* (Adam Hilger, Ltd., Techno House, 1990)
15. M. Born, E. Wolf, *Principles of Optics*, 2nd ed., (Pergamon Press Ltd., Oxford, 1964), pp. 328
16. A.M. Hamed, Topics on optical and digital image processing using holography and speckle techniques, publisher by Lulu.com, ISBN 9781329328464 Nov. 29 (2015)

Thesis

17. A.M Hamed, Some applications of scattered light and multiple beam interference using coherent light, M.Sc. Thesis (1976)

Chapter 3
Step Index Fiber Using Multiple Laser Beam Interferometers

A model is suggested to describe the fringe shift that occurred due to the phase variations of the clad glass fiber introduced between the two plates of the liquid wedge interferometer illuminated with a He–Ne laser.

The fringe shift of the phase object is represented in the harmonic term, which appears in the denominator of the Airy distribution formula for multiple beams of the Fabry and Perot interferometer. An experiment is conducted using a liquid wedge interferometer where a step-index glass fiber with a nearly quadratic thickness variation is introduced between the two plates of the interferometer. The obtained fringe shift shows good agreement with the proposed quadratic model. The MATLAB code is written to plot the interferometer fringes comprising the shift of the step-index fiber. Second, the recognition of elliptical fibers is outlined using tomographic imaging. Finally, the concluding results are given.

3.1 Introduction

TOLANSKY [1] obtained an approximate formula of Airy summation and noted the main features between the beams forming multiple beam fringes at infinity by using a plane parallel plate and those forming multiple beam localized fringes. In the case of the wedge, the successively multiplied reflected beams are not in phase in the exact arithmetic series, while in the case of plane-parallel plates, the path difference between any two successive beams is $\lambda/2$. The optimum condition for producing multiple localized beam fringes, reached by TOLANSKY, necessitates a small interferometer wedge angle α to fulfill Airy's summation conditions. Barakat and Mokhtar [2] reported that the permitted limit is 3/8 λ. In another study [3–10], interference was obtained using synthetic optical fibers. They considered ray optics approximations using monochromatic light emitted from mercury and other spectral lamps. Additionally, Barakat obtained the right formula, which is based on ray optics,

© The Author(s), under exclusive license to Springer Nature Switzerland AG 2024   27
A. Hamed, *Cascaded Interferometers and Their Medical Applications*,
SpringerBriefs in Applied Sciences and Technology,
https://doi.org/10.1007/978-3-031-64535-8_3

of multiple beam fringes crossing a fiber of circular transverse cross section immersed in a silvered liquid wedge. His work was followed by that of others [8], who extended the analysis to multilayer fibers. Hamza et al. [9] determined the refractive indices and birefringence of fibers with irregular transverse sections of homogeneous fibers. Boggs et al. [11] and Presby et al. [12] described the automated transverse inter- ferometer method and described the index profile of a graded index fiber. Recently, Hamed [13] considered the effect of the wedge angle α on the arithmetic series using Gaussian laser illumination. The modified Airy distribution is obtained using Fourier imaging of fibers. The effect of laser modulation on the contrast and sharpness of fringes has been investigated. Then, the fringe shift of unclad fibers is experimentally obtained, and a model is constructed to describe this single shift [14].

The radon function computes *projections* of an image matrix along specified directions. A projection of a two-dimensional function is a set of line integrals. The radon function computes the line integrals from multiple sources along parallel paths, or *beams*, in a certain direction. The beams are spaced 1-pixel apart. To represent an image, the radon function takes multiple parallel-beam projections of the image from different angles by rotating the source around the center of the image [15–19]. Three-dimensional (3D) imaging by holographic tomography can be performed for a fixed detector through the rotation of either the object or the illumination beam [20].

In this chapter, we suggest a model of two-step variations to describe the thickness distribution of step-index fibers with circular transverse cross sections. The Airy distribution is written in the case of fiber modulation, giving nearly quadratic shift variations inside the core.

An experiment [5] was performed using a liquid wedge interferometer where a step-index fiber was enclosed between the two plates of the interferometer. The interferometer is illuminated by a He-Ne laser. The quadratic theoretical model is compared with the experimental shift of the fiber, which shows good agreement. In the following section, a theoretical analysis is presented including the quadratic model for the fringe shift. Finally, the results and discussion are given.

3.2   Theoretical Analysis

3.2.1   Model of the Fringe Shift in Clad Fiber

We propose the following model to describe a clad fiber placed inside a liquid wedge interferometer. For simplicity, we assume square interferometer plates of dimensions 2a, and 2b, and a refractive index μ_L. The fiber radius is r_f, and the core radius is r_c with skin and core indices μ_s, and μ_c respectively.

Hence, an object immersed in a liquid is analytically represented as follows:

$$G(x, y) = g_c(x, y) + g_s(x, y) + g_L(x, y) \tag{3.1}$$

$g_c(x, y) = 1$; for $x^2 + y^2 < r_c^2$ for the core central region.

$g_s(x, y) = 1$; for $r_c^2 < x^2 + y^2 < r_f^2$ for the skin region of the fiber.

$g_L(x, y) = 1$; for $x < (a - r_f)$ and $y < (b - r_f)$ with $a = b$ for a square liquid wedge.

For uniform illumination emitted from the spatially filtered laser beam, the Fourier spectrum of the object information is calculated by operating the F.T. with Eq. (3.1) to obtain:

$$\tilde{g}(u, v) = \frac{2J_1(w_1)}{w_1} + \left[\frac{2J_1(w_2)}{w_2} - \frac{2J_1(w_1)}{w_1} \right] + \frac{\sin(x)}{x} \frac{\sin(y)}{y} \tag{3.2}$$

where $\rho = (u, v)$ is the radial coordinate in the Fourier plane and $r = (x, y)$ is the radial coordinate in the object plane. The reduced coordinates W_1 and W_2 are given by:

$$W_1 = 2\pi \frac{r_c \rho}{\lambda f}, \text{ and } W_2 = 2\pi \frac{r_f \rho}{\lambda f}$$

A quadratic model is assumed for the fringe shift of the step-index fiber. It is mathematically represented as follows:

$$t_p(y) = t_0 \left(\frac{r_c}{r_f} \right) \left[1 - \left(\frac{y_1}{r_c} \right) \right]^2 + t_0 \left(\frac{r_s}{r_f} \right) \left[1 - \left(\frac{y_2}{r_s} \right) \right]^2; 0 < y_1 < r_c \text{ and } r_c < y_2 < r_f$$

$$= t; \quad y > y_1 + y_2 \tag{3.3}$$

In Eq. (3.3), t_0 is the maximum shift inside the fiber.

The fringe phase shift in a liquid wedge interferometer provided with a step-index fiber is described as follows:

The phase of the step-index fiber is calculated as follows [21]:

$$\phi(y) = \left(\frac{2\pi}{\lambda} \right) \left[2\mu_L t + 4(\mu_c - \mu_L) t_0 \left(\frac{r_c}{r_f} \right) \right] \left[1 - \left(\frac{y_1}{r_c} \right) \right]^2$$

$$+ 4(\mu_s - \mu_L) t_0 \left(\frac{r_s}{r_f} \right) \left[1 - \left(\frac{y_2}{r_s} \right) \right]^2 \tag{3.4}$$

Since $t = z \tan\alpha$ for the wedge interferometer of wedge angle (α), Eq. (3.4) becomes:

$$\phi(y) = \left(\frac{2\pi}{\lambda} \right) \left[2\mu_L z \tan(\alpha) + 4(\mu_c - \mu_L) t_0 \left(\frac{r_c}{r_f} \right) \right] \left[1 - \left(\frac{y_1}{r_c} \right) \right]^2$$

$$+ 4(\mu_s - \mu_L) t_0 \left(\frac{r_s}{r_f} \right) \left[1 - \left(\frac{y_2}{r_s} \right) \right]^2 \tag{3.5}$$

The interfringe spacing of the straight-line fringes is given by $\Delta z = \lambda/2$.

3.2.2 Tomographic Imaging of Different Cross Sections of Fibers

The circular cross section of the unclad fiber is defined in the (x, y) plane as follows:

$$P(x, y) = 1 \; ; \; |r/r_f| \leq 1 \tag{3.6}$$

In Eq. (3.1), rf is the radius of the fiber.

Assuming Gaussian illumination from the laser beam, the whole tomography pattern is built by scanning the object at different angles from $0°$ to $180°$.

For the image recorded at a certain angle θ, Fig. 3.1, is given as follows:

$$R_\theta(x')_{gauss} = \int_{-\infty}^{\infty} \exp[-(x^{2'} + y^{2'})/w_o^2]$$

$$P(x' \cos\theta - y' \sin\theta, x' \sin\theta + y' \cos\theta) \; dy' \tag{3.7}$$

Since there is a radial symmetry of revolution for the circular cross section of the fiber, $r(x, y) = r(x', y')$, and w_0 is the waist of the laser beam.

For coherent uniform illumination incident upon the aperture, the complex amplitude of the tomography pattern is computed as follows:

$$R_\theta(x')_{uniform} = \int_{-\infty}^{\infty} P(x' \cos\theta - y' \sin\theta, x' \sin\theta + y' \cos\theta) \; dy' \tag{3.8}$$

The point spread function (PSF) of the imaging system, using Gaussian illumination, is computed by applying the Fourier transform to the multiplication product of the fiber cross section and the Gaussian illumination function as follows:

$$PSF = F.T.\{\exp[-(x^2 + y^2)/w_o^2].P(x, y)\}$$

$$= \exp[-w_o^2(u^2 + v^2)] \otimes 2J_1(u^2 + v^2)/(u^2 + v^2) \tag{3.9}$$

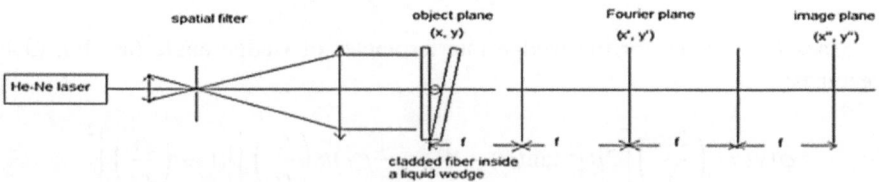

Fig. 3.1 Optical system used to record the modulated multiple beam interference using a clad fiber as an object. The liquid wedge interferometer has a refractive index of 1.516, while the fiber refractive indices are 1.5154 for the core and 1.5161 for the skin

Symbol $\otimes$ in Eq. (3.9) is used for the convolution operation. Hence, the PSF of the imaging system is computed from the convolution product of the Fourier transform of each function, as shown in Eq. (3.10).

$$F.T.[-(x^2 + y^2)/w_o^2] = \exp[-w_o^2(u^2 + v^2)]$$

$$and \quad F.T.[P_1(x, y)] = \frac{2J_1[\alpha(u^2 + v^2)^{1/2}]}{[(u^2 + v^2)^{1/2}\alpha]}; \ \alpha = 2\pi r_f/\lambda f \qquad (3.10)$$

The point spread function (PSF) is computed numerically by taking the Fourier transform of the fiber cross section.

In the case of a step index fiber, the PSF of the circular cross section is computed as follows:

$$F.T.\left[P_2(x, y)\right] = \frac{2J_1[\alpha_1(u^2 + v^2)^{1/2}]}{[(u^2 + v^2)^{1/2}\alpha_1]} - \frac{2J_1[\alpha_2(u^2 + v^2)^{1/2}]}{[(u^2 + v^2)^{1/2}\alpha_2]};$$

$$\alpha_1 = 2\pi r_f/\lambda f \quad and \quad \alpha_2 = 2\pi r_c/\lambda f \qquad (3.11)$$

3.3 Results and Discussion

The optical system used to record multiple beam interference modulated by a clad glass fiber is shown in Fig. 3.1. The system is illuminated by a He-Ne laser at $\lambda = 633$ nm and is spatially filtered before being incident upon the interferometer jig. The interference is imaged using an optical microscope. The liquid wedge interferometer has a refractive index of 1.516, while the fiber refractive indices are 1.5154 for the core and 1.5161 for the skin.

The theoretical quadratic model is drawn using Eq. (3.5) and plotted as shown in Figs. 3.2, 3.3. Theoretical imaging of multiple beam interference modulated by a step index fiber with a fiber radius equal to two times the core radius. The image in Fig. 3.2 has matrix dimensions of 1024×1024 pixels, the fiber diameter is 128 pixels, and the core diameter is 64 pixels. Figure 3.3 shows a theoretical image of multiple beam interference modulated by a step index fiber with a fiber radius four times the core radius. Additionally, the image has matrix dimensions of 1024×1024 pixels, the fiber diameter is 128 pixels, and the core diameter is 32 pixels.

An experimental photograph of the fringe shift that occurred due to multiple beam interference from the clad fibers is shown in Fig. 3.4. The total fiber diameter is $t_f = 201.21 \ \mu$m. The skin refractive index $= 1.5154$, the core refractive index $= 1.5161$, and the liquid refractive index $= 1.516$. The wavelength of the He-Ne laser $=$ was 632.8 nm. The differential fringe shift of the core $=$ is 0.1931. The experiment was carried out at room temperature at $T = 32$ °C. The theoretical model of the fringe shift of the clad fibers, as shown in Figs. 3.2, 3.3 agreed well with the experimental fringe shift for the clad glass fibers, as shown in Fig. 3.4.

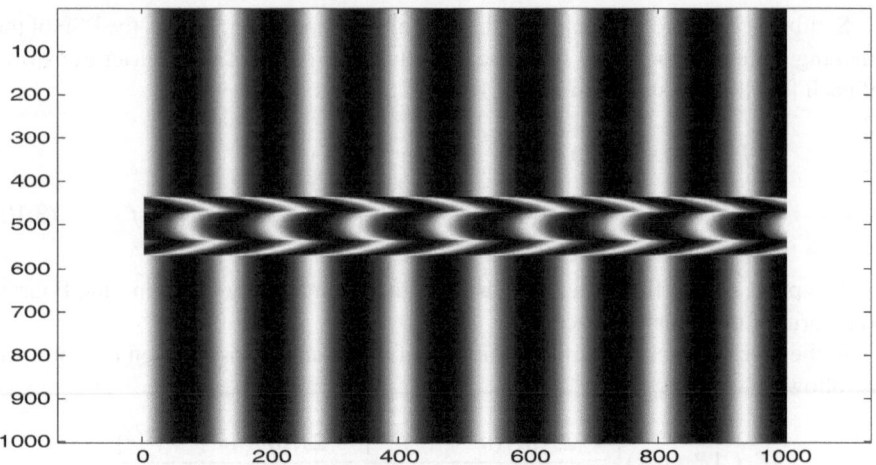

Fig. 3.2 Theoretical imaging of multiple beam interference modulated by a step index fiber with a fiber radius equal to two times the core radius. The image has matrix dimensions of 1024 × 1024 pixels. The fiber diameter is 128 pixels, while the core diameter is 64 pixels

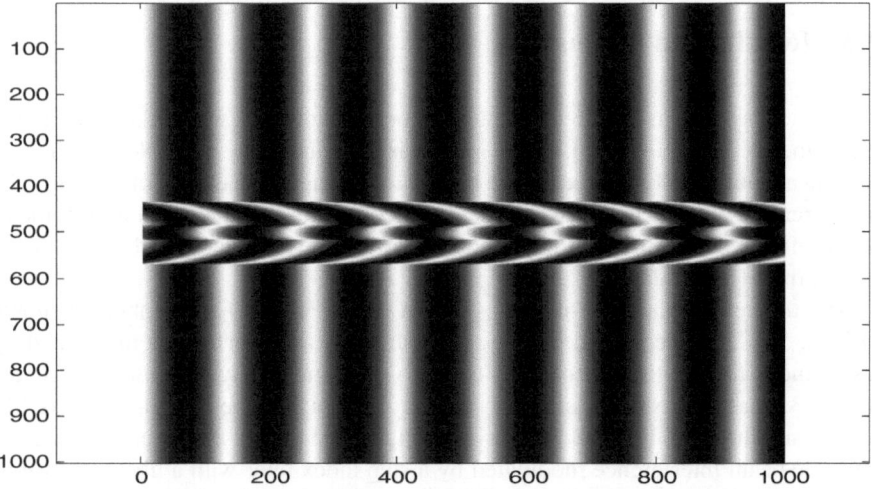

Fig. 3.3 Theoretical imaging of multiple beam interference modulated by a step index fiber with a fiber radius equal to four times the core radius. The image has matrix dimensions of 1024 × 1024 pixels. The fiber diameter is 128 pixels, while the core diameter is 32 pixels

The tomographic images of the different circular cross sections of the unclad, step index, and graded index fibers are plotted in Fig. 3.5. The tomographic image for the unclad fiber shows a solid cylinder, while the step index and graded index fibers have a cylinder of graded illumination. The elliptic deformation from the uniform circular cross section is illustrated in Fig. 3.6.

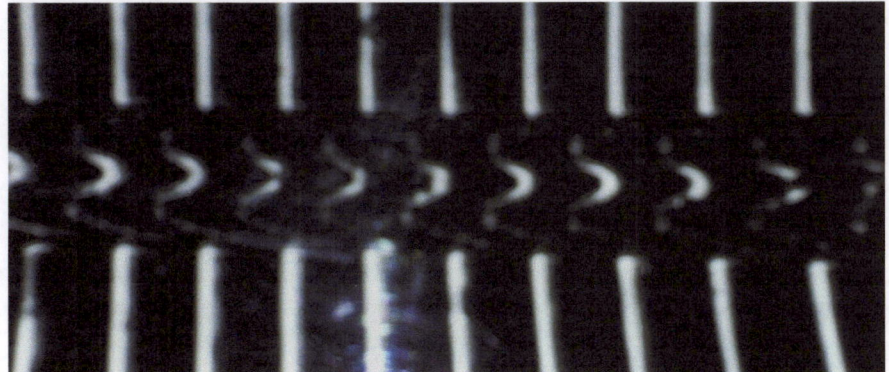

Fig. 3.4 Experimental photograph of straight-line fringes modulated by the shift introduced by the clad glass fiber. The total fiber diameter is $r_f = 201.21$ µm. The skin refractive index $= 1.5154$, the core refractive index $= 1.5161$, and the liquid refractive index $= 1.516$. The wavelength of the He–Ne laser $= 6328$ µm. The differential fringe shift of the core $= 0.1931$. The experiment was carried out at room temperature at $T = 32$ °C

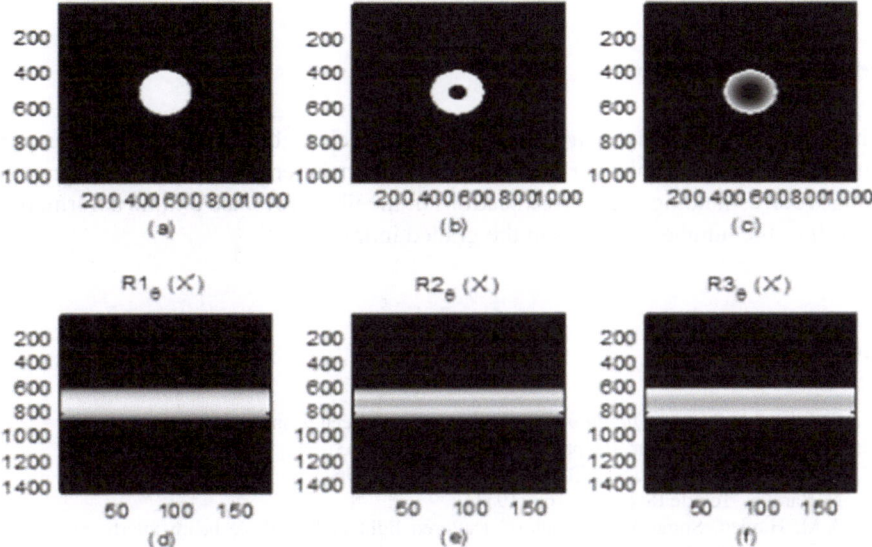

Fig. 3.5 a Cross-sectional view of a uniform circular fiber: **a** an unclad fiber, **b** a step-index fiber, and **c** a graded index fiber, the corresponding tomographic images are shown in **d**, **e**, and **f**

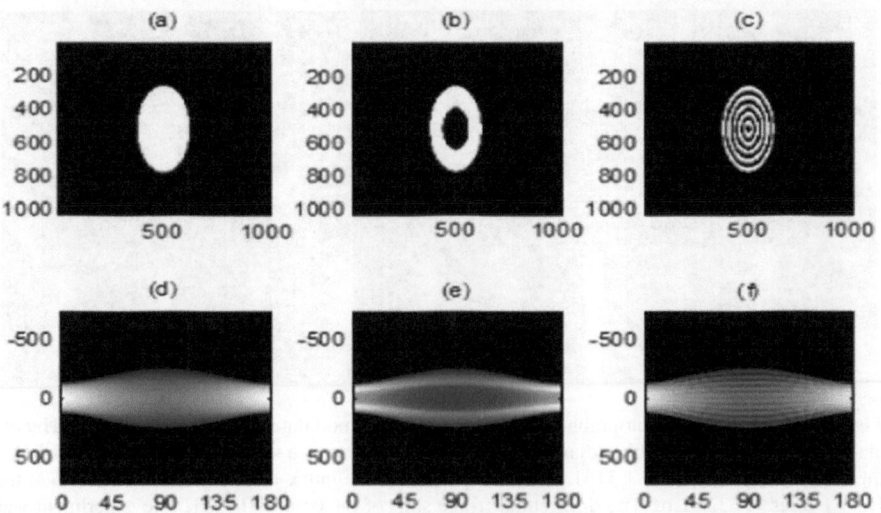

Fig. 3.6 **a** Cross-sectional view of an ellipse: **a** an unclad fiber, **b** a step-index fiber, and **c** multistep index fiber, the corresponding tomographic images are shown in **d**, **e**, and **f**

3.4 Conclusion

The proposed theoretical quadratic model for the clad fibers agrees with the experimental fringe shift obtained under multiple beam interference. The tomographic images corresponding to the cross sections of the fibers revealed elliptic deformation as well as the number of layers in the graded index fibers.

References

1. S. Tolansky, *Surface Microtopography* (Longmans Green, London, 1960)
2. N. Barakat, S. Mokhtar, J. Opt. Soc. Am. **53**, 159 (1963)
3. N. Barakat, A.M. Hindelah, Textile Res. J. **41**, 581 (1964)
4. N. Barakat, Textile Res. J. **41**, 167 (1971)
5. A.M. Hamed, Some applications of scattered light and multiple beam interference using coherent light, M.Sc. Thesis (1976)
6. N. Barakat, A. Hamza, *Interferometry of Fibrous Materials* (Adam Hilger, Ltd. Techno House, Bristol, 1990)
7. M.M. El Nicklawy, I.M. Fouda, Textile Res. J. **71**, 252 (1980)
8. A. Hamza, T.Z.N. Sokkar, M.A. Abeel, J. Phys. D. **18**, 1773 (1985)
9. A. Hamza, T.Z.N. Sokkar, M.A. Abeel, J. Phys. D. **19**, 119 (1986)
10. N. Barakat, A. Hamza, A. Goned, Appl. Opt. **24**, 4383 (1985)
11. L.M. Boggs, H.M. Presby, D. Marcuse, Bell System Tech. J. **58**, 867 (1979)
12. H.M. Presby, D. Marcuse, H.W. Astle, L.M. Boggs, Bell Syst. Tech. J. **58**, 883 (1979)
13. A.M. Hamed, Opt. Appl. **27**, 229 (1997)
14. A.M. Hamed, Pramana. J. Phys. **70**, 643 (2008)

15. R. Koprowski, Z. Wróbel, *Image Processing in Optical Coherence Tomography using Mat-lab, Poland* (Univ, Silesia, 2011)
16. R. Gonzalez, R. Woods, *Digital Image Processing* (Addison-Wesley Publishing Company, 1992)
17. M. Akiba, K.P. Chan, N. Tanno, Full-field optical coherence tomography by two-dimensional heterodyne detection with a pair of CCD cameras. Optics Lett. **28** (2003)
18. D.C. Adler, T.H. Ko, J.G. Fujimoto, Speckle reduction in optical coherence tomography images by use of a spatially adaptive wavelet filter. Opt. Lett. **29** (2004)
19. K. Vandersteen, B. Busselen, K. Van Den Abeele, J. Carmeliet, Quantitative characterization of fracture apertures using microfocus computed tomography. Geol. Soc. Spec. Pub. **215**, 61–68 (2003)
20. S.S. Kou, J.R.C. Sheppard, Image formation in holographic tomography: high-aperture imaging. Appl. Opt. **48**, 168–175 (2009)
21. A.M. Hamed, Pramana J. Phys. **82**(3), 529–536 (2014)

Chapter 4
Modeling of the Fringe Shift for Unclad Glass Fibers Using Ordinary Multiple Beam Interference

A quadratic model is suggested to describe the fringe shift that occurred due to the phase variations of unclad glass fibers introduced between the two plates of the liquid wedge interferometer. The fringe shift of the phase object is represented in the harmonic term, which appears in the denominator of the Airy distribution formula of Fabry–Perot's interferometer. A computer program is written to plot the computer fringe shifts of the described model.

An experiment is conducted using a liquid wedge interferometer where a fiber with a nearly quadratic thickness variation is introduced between the two plates of the interferometer. The obtained fringe shift shows good agreement with the proposed quadratic model. Additionally, it is compared with the previous theoretical shift based on ray optics of semicircular shape.

4.1 Introduction

TOLANSKY [1] obtained an approximate formula of Airy summation and noted the main features between the beams forming multiple beam fringes at infinity by using a plane parallel plate and those forming multiple beam localized fringes. In the case of the wedge, the successively multiple reflected beams are not in phase in the exact arithmetic series, while in the case of plane parallel plates, the phase difference between any two successive beams is $\lambda/2$. The optimum condition for producing multiple localized beam fringes, reached by TOLANSKY, necessitates a small interferometric wedge angle α to fulfill Airy's summation conditions.

Barakat and Mokhtar [2] reported that the permitted limit is (3/8) λ. In another study [3–10], interference was obtained using synthetic optical fibers. They considered ray optics approximations using monochromatic light emitted from mercury and other spectral lamps.

© The Author(s), under exclusive license to Springer Nature Switzerland AG 2024   37
A. Hamed, *Cascaded Interferometers and Their Medical Applications*,
SpringerBriefs in Applied Sciences and Technology,
https://doi.org/10.1007/978-3-031-64535-8_4

Additionally, Barakat obtained the right formula, which is based on ray optics, of multiple beam Fizeau fringes crossing a fiber of circular transverse cross section immersed in a silvered liquid wedge. His work was followed by that of others [8], who extended the analysis to multilayer fibers. Hamza et al. [9] determined the refractive indices and birefringence of fibers with irregular transverse sections of homogeneous fibers. Boggs et al. [11] and Presby et al. [12] described the automated transverse interferometric method and described the index profile of graded index fibers. Recently, Hamed [13] considered the effect of the wedge angle α on the arithmetic series using Gaussian laser illumination. The modified Airy distribution is obtained using Fourier imaging of fibers. The effect of laser modulation on the contrast and sharpness of fringes has been investigated.

In this chapter, we suggest a model of quadratic variations to describe the thickness variations of fibers with circular transverse cross sections. The Airy distribution is written in the case of fiber modulation giving nearly quadratic shift variations.

An experiment [5] is performed using a liquid wedge interferometer where a fiber is enclosed between the two plates of the interferometer. The interferometer is illuminated by a He–Ne laser. The quadratic theoretical model is compared with the experimental shift of the fiber, which shows good agreement. In the following section, a theoretical analysis is presented including the quadratic model for the fringe shift. Finally, the results and discussion are given.

4.2 Theoretical Analysis

A collimated laser beam emitted from the He–Ne laser at $\lambda = 632.8$ nm is used to illuminate the wedge interferometer, as shown in Fig. 4.1. The interferometer is formed by enclosing a matched liquid of suitable refractive index (μ_L) situated between two silvered optical plates forming a wedge gap (α). Hence, the examined phase object is located between the two optical plates. The recorded transmitted intensity of multiple beam interference is represented by the well-known expression Airy distribution [1–5], where $k = 2\pi/\lambda$ is the propagation constant, r_1 is the amplitude reflection coefficient of either of the interferometer plates, $r \equiv (x, y)$ is the imaging plane where the fringes are located, and Δ is the optical path difference between the first two transmitted rays, with ε being the differential increment introduced in the successive transmitted rays.

From the theory of Fizeau fringes applied to fibers [4, 5], using the ray optics approximation, we obtain:

$$I(r) = I_o \frac{1}{1 + F \sin^2(\frac{\delta}{2})} \tag{4.1}$$

$$(\mu_L - \mu_f)t_f = \frac{dz}{\Delta Z}(\lambda/2) \tag{4.2}$$

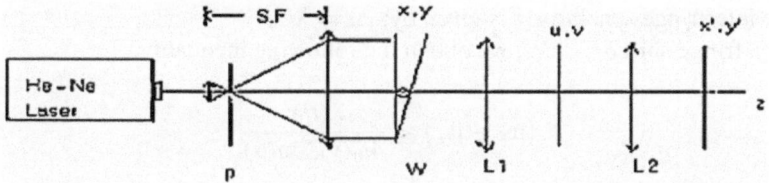

Fig. 4.1 Imaging system of multiple beam Fizeau fringes using a liquid wedge interferometer (W) located in the plane (x, y), where SF is a spatial filter for rendering the beam parallel to a certain cross section with a pinhole p having a diameter sufficient to pass through the first order of the diffraction pattern. L_1 and L_2 are Fourier transform lenses, each with a focal length of 20 cm (u, v) is the spectral Fourier plane of the object plane (x, y). (x, y) is the imaging plane where the fringes are formed

where $t_f = 2r_f$ is the fiber diameter of the refractive index μ_f, $\Delta z = \lambda/2$ is the distance between any two successive straight-line fringes and dz is the fringe shift introduced due to the fiber. In Eq. (4.1), F is the coefficient of finesse and δ is the optical phase difference.

For the transverse section of an unclad fiber with a constant refractive index μ_f the variable thickness t_f varies quadratically, reaching the maximum value at the center of the fiber. Hence, we propose a model to describe the shift introduced by the fiber and compare it with the experimental shift. Additionally, it is compared with the theoretical shift that is based on ray optics. The parabolic function is introduced for the fringe shift of width equal to that of the fiber diameter and is represented in the following paragraph.

4.3 A *Parabolic* Model for the Fringe Shift

A parabolic function is assumed to represent the fringe shift obtained in the case of multiple beam interference occurring in a liquid wedge interferometer. This means that the utilization of the phase object of thickness $t_p(y)$ varies in a quadratic parabolic shape, keeping the refractive index fixed at μ_f. In this case, the quadratic function is represented as follows:

$$t_p(y) = t_0(\frac{y}{t_f})^2; y \le t_f$$
$$= t; y > t_f \tag{4.3}$$

where t_0 is the maximum shift.

The phase of the object is calculated as follows:

$$\phi(y) = \frac{2\pi}{\lambda} \left[2\mu_L t + 4t_p(y)(\mu_f - \mu_L)\right]$$
$$= \frac{2\pi}{\lambda} \left[2\mu_L Z \tan(\alpha) + 4t_0(\frac{y}{t_f})^2(\mu_f - \mu_L)\right] \tag{4.4}$$

The interfringe spacing Δz is given by: $\Delta z = \lambda/2$.
For a fringe shift $\delta z < \Delta z$, we obtain the following inequality:

$$(\mu_f - \mu_L) < \frac{\lambda (t_f)^2}{4 t_0 (y)^2 \sin(\alpha)} \tag{4.5}$$

Additionally, no shift occurs when $\mu_f = \mu_L$.

The straight-line fringes modulated by the quadratic function are represented by the following formula:

$$m\lambda = 2\mu_L Z \tan(\alpha) + 4 t_0 (\mu_f - \mu_L)(\frac{y}{t_f})^2 \tag{4.6}$$

The differential fringe shift is calculated as follows:

$$D_z(y) = \frac{dz}{\Delta z} = \frac{4 t_0 (\mu_f - \mu_L)(\frac{y}{t_f})^2}{\lambda \sin(\alpha/2)} \tag{4.7}$$

Hence, $D_z(y)$ represents an exact quadratic shift for the cited model.

4.4   Results and Discussion

An experimental photograph of the straight-line fringes modulated by a shift due to the introduction of an unclad glass fiber between the two plates of the Fabry–Perot interferometer is shown in Fig. 4.2. This photo is obtained in the imaging plane (x, y), as shown in Fig. 4.1. The arrangement is considered a liquid wedge interferometer illuminated with a collimated beam emitted from a He–Ne laser at $\lambda = 632.8$ nm. The refractive index of the fiber $\mu_f = 1.5172$ is greater than the refractive index of the liquid $\mu_L = 1.5158$. The differential fringe shift is computed as $dz/z = 0.4907$ using a traveling microscope with an accuracy of $10\,\mu$m. Figure 4.3a–d represent the plot of fringes modulated by the parabolic function computed from Eq. (4.6). The differential fringe shift is taken as $dz/\Delta z = 0.25, 0.5, 1$, and 1.25, and the full width is set equal to the fiber diameter at $t_f = 109\,\mu m$. A comparative plot represented by a semicircular function with a width equal to that of the fiber diameter at $t_f = 109\,\mu m$ is shown in Fig. 4.4. The quadratic model shows good agreement with both the experimental photograph and the semicircular shape. A MATLAB program is written to plot Fig. 4.3a–d using Eq. (4.6). Other experimental results of the fringe shift at different diameters and different refractive indices are presented in [5].

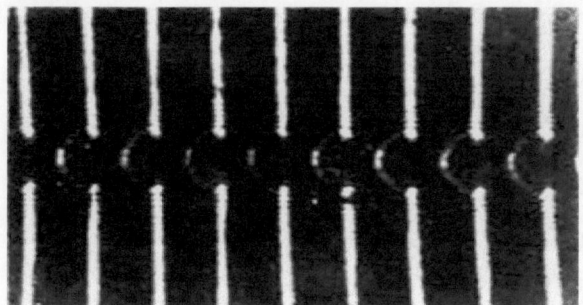

Fig. 4.2 Experimental photograph of straight-line fringes modulated by the phase shift introduced by the fiber, where $d = 109.03$ μm is the fiber diameter, $\mu_L = 1.5154$ is the liquid refractive index, $\mu_{fiber} = 1.5172$ is the fiber refractive index, $dz/z = 0.4907$ is the differential fringe shift, and $\lambda = 632.8$ nm is the laser wavelength. The experiment was performed at room temperature $T = 25.5$ °C

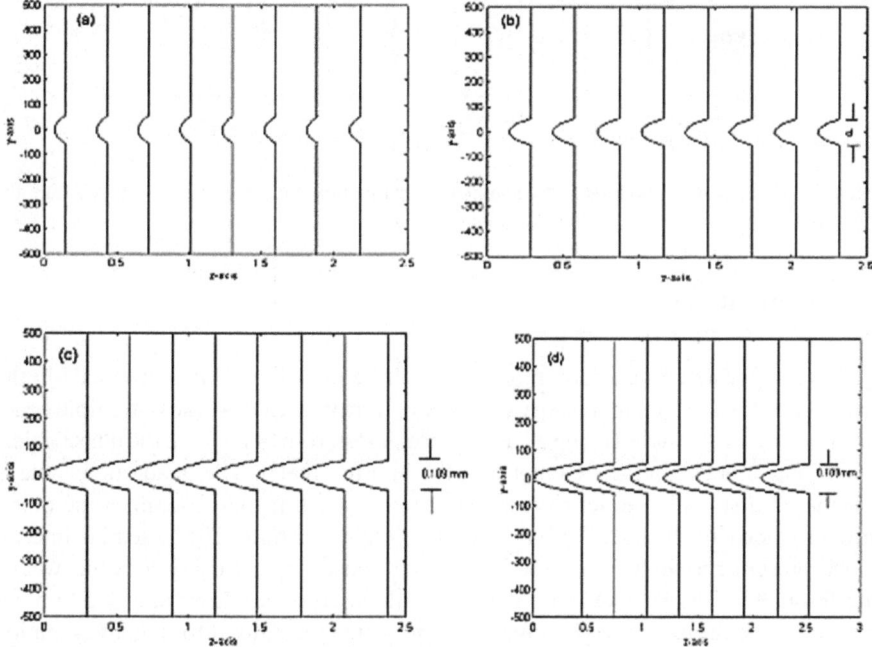

Fig. 4.3 Plots of fringes modulated by a parabolic function with full width $= 0.109$ mm. The y-axis and z-axis are on the micrometric scale, and the differential fringe shifts are equal to **a** 0.25, **b** 0.5, **c** 1.0, and **d** 1.25

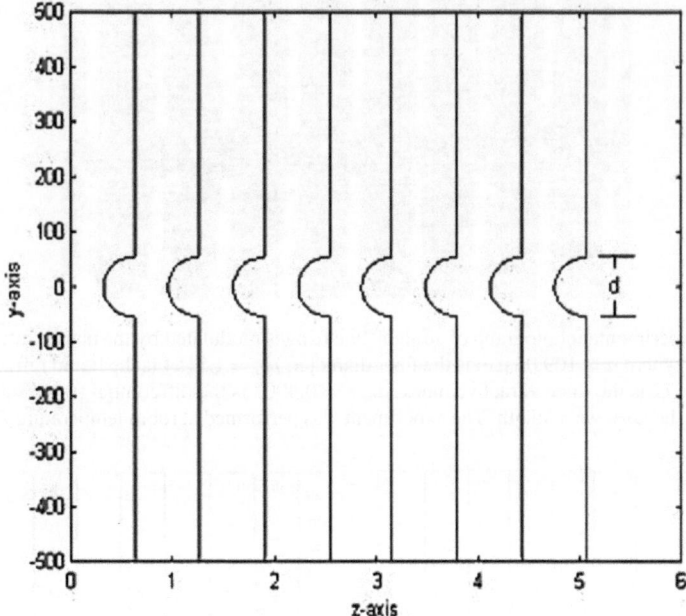

Fig. 4.4 Plot of fringes modulated by a semicircular function with diameter $d = 0.109$ mm, the y-axis and z-axis are on the micrometric scale

4.5 Conclusion

We have suggested a quadratic model to describe the fringe shift introduced in the phase term that appears in the Airy distribution formula. In this study, multiple beam interference is considered using a Fabry–Perot interferometer, where the phase object is introduced between the two plates of the interferometer. The quadratic profile is considered a suitable model to represent the fringe shift that resembles the phase shift produced by fibers of the transverse circular section. The potential interest of the present research is to process colored phase objects using a liquid wedge interferometer. The He–Ne laser is mixed with an argon ion laser at $\lambda = 514.5$ nm to produce a collimated polychromatic beam, which is required for the investigation of colored phase objects.

References

1. S. Tolansky, *Surface Microtopography* (Longmans Green, London, 1960)
2. N. Barakat, S. Mokhtar, J. Opt. Soc. Am. **53**, 159 (1963)
3. N. Barakat, A.M. Hindelah, Textile Res. J. **41**, 581 (1964)
4. N. Barakat, Textile Res. J. **41**, 167 (1971)

5. A.M. Hamed, Some applications of scattered light and multiple beam interference using coherent light, M.Sc. Thesis (1976)
6. N. Barakat, A. Hamza, *Interferometry of Fibrous Materials* (Adam Hilger, Ltd., Techno House, Bristol, 1990)
7. M.M. El-Nicklawy, I.M. Fouda, Textile Res. J. **71**, 252 (1980)
8. A. Hamza, T.Z.N. Sokkar, M.A. Abeel, J. Phys. D **18**, 1773 (1985)
9. A. Hamza, T.Z.N. Sokkar, M.A. Abeel, J. Phys. D **19**, 119 (1986)
10. N. Barakat, A. Hamza, A. Goned, Appl. Opt. **24**, 4383 (1985)
11. L.M. Boggs, H.M. Presby, D. Marcuse, Bell Syst. Tech. J. **58**, 867 (1979)
12. H.M. Presby, D. Marcuse et al., Bell Syst. Tech. J. **58**, 883 (1979)
13. A.M. Hamed, Fourier imaging of uncladded fibers using a liquid wedge interferometer. Opt. Applicate **27**, 229–240 (1997)

Chapter 5
Recognition of Some Modulated Apertures Using the Cascaded Fabry–Perot Interferometer (CFPI)

In this chapter, five modulated apertures are considered. The first has a linear distribution, the second has a conic shape, the third has a quadratic distribution, the fourth has a Gaussian distribution, and the fifth has B/W concentric annuli. These apertures are recognized from the fringe shift that occurred in the interferometric images using the CFPI. The recognized modulated apertures are compared with the fringe shift corresponding to the uniform circular aperture. An improved multiple beam interferometer or a CFPI working in transmission is considered for aperture recognition. The sharp interferometric images obtained from the CFPI provided with the apertures are constructed using MATLAB code.

5.1   Introduction

An early digital speckle pattern interferometric system for monitoring surface vibrations and out-of-plane tilt was presented [1], where the resolution of the system used to measure the out-of-plane displacement was $\lambda/2$ per fringe. The application of a general-purpose image-processing computer system for automatic fringe analysis is presented. The applications considered are strain measurement by speckle interferometry, position location in three axes, and fault detection in holographic nondestructive testing [2].

Recently, image processing of apertures using speckle photography and holography has been studied by many authors [3–7]. Interferometric microscopy and other images were investigated in [8–15]. In a recent publication by the author [16], multiple passes of two-beam interference were considered, while in this study, we considered CFPI for the formation of sharp fringes.

In this chapter, aperture recognition using a cascaded Fabry–Perot interferometer (CFPI) is proposed. We can recognize the shape of the apertures from the fringe shift. Modulated apertures such as linear, quadratic, B/W concentric annulus, and Gaussian

© The Author(s), under exclusive license to Springer Nature Switzerland AG 2024   45
A. Hamed, *Cascaded Interferometers and Their Medical Applications*,
SpringerBriefs in Applied Sciences and Technology,
https://doi.org/10.1007/978-3-031-64535-8_5

distributions are designed using the MATLAB code. The results and discussion are given followed by a conclusion.

5.2 Theoretical Analysis

A higher-order multiple beam interference composed of four cascaded interferometers is schematically represented in Fig. 5.1. A He-Ne laser beam is spatially filtered and rendered parallel using an objective lens L followed by a pinhole P placed in the short focus of the objective lens, and a converging lens L_1 placed at the focal plane f from the pinhole. The collimated laser beam passes through the four F.P.I.s arranged in series followed by the Fourier transform lens L_2 of focal length f_2. The Fourier and imaging planes are located as shown in Fig. 5.1.

The intensity distribution in the case of an ordinary FPI is given by the following formula [13]:

$$I(\delta; R) = \frac{T^2}{1 + R^2 - 2R\cos(\delta)} \tag{5.1}$$

where T is the transmission efficient and R is the reflection coefficient of the interferometer. δ, is the phase difference between any two adjacent emerging rays.

In the case of cascaded interferometers, the intensity distribution is an ordinary distribution to a power of N, where N is the number of cascaded interferometers. Then, the intensity distribution in the cascaded Fabry–Perot interferometer (CFPI) is represented as follows:

$$I(\delta; R, N) = \left[\frac{T^2}{1 + R^2 - 2R\cos(\delta)} \right]^N \tag{5.2}$$

The maximum intensity is computed as:

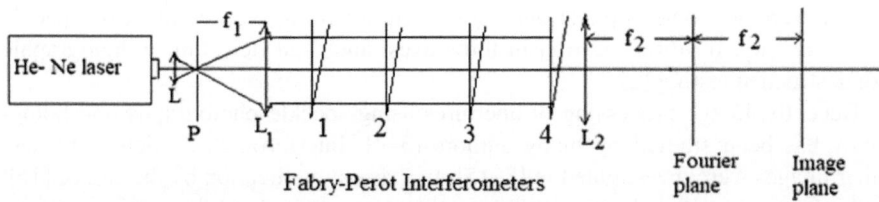

Fig. 5.1 Multiple beam interferometer composed of four cascaded interferometers. L objective lens, P pinhole, and L_1 converging lens where the elements L, P, and L_1 render the laser beam spatially filtered and collimated. L_2 is the Fourier transform lens of focal length f_2. The Fourier and imaging planes are shown in the figure

$$I(\delta = 2\pi; R, N) = I_{\max}(R, N) = \left[\frac{T^2}{(1-R)^2}\right]^N \tag{5.3}$$

Using Eqs. (5.2, 5.3), the normalized intensity due to cascaded multiple beam interference can be written as follows:

$$I_{\text{normalized}}(\delta; R, N) = \frac{I}{I_{\max}} = \frac{1}{\left[1 + F \sin^2(\frac{\delta}{2})\right]^N} \tag{5.4}$$

Define

$$F' = (1 + F)^N - 1 \tag{5.5}$$

where $F = \frac{4R}{(1-R)^2}$ defined as the coefficient of finesse for ordinary FPI is a measure of fringe sharpness and contrast and F' is the coefficient of finesse for the CFPI.

For ordinary FPI, $N = 1$ and we obtain: $F' = F$.

For greater values of reflectivity, $R > 80\%$, F is much larger than one, hence, an approximate expression for the intensity is obtained as:

$$I_{\text{normalized}}(\delta; R, N) \sim 1/[F \sin^2\left(\frac{\delta}{2}\right)]^N \tag{5.6}$$

The five modulated apertures used in the CFPI described above are represented as follows:

$$P(\rho) = \rho; \left|\frac{\rho}{\rho_0}\right| \leq 1; \text{ for linear aperture} \tag{5.7}$$

$$P(\rho) = 1 - \rho; \left|\frac{\rho}{\rho_0}\right| \leq 1; \text{ for conic aperture} \tag{5.8}$$

$$P(\rho) = \rho^2 ; \left|\frac{\rho}{\rho_0}\right| \leq 1; \text{ for quadratic aperture} \tag{5.9}$$

$$P(\rho) = \exp[-\left(\frac{\rho}{\rho_0}\right)^2]; \left|\frac{\rho}{\rho_0}\right| \leq 1; \text{ for Gaussian aperture} \tag{5.10}$$

$$P(\rho) = \sum_{i=1}^{N}[P_{2i}(\rho) - P_i(\rho)]; \left|\frac{\rho}{\rho_0}\right| \leq 1 \text{ for black and white concentric annuli} \tag{5.11}$$

The apertures described in Eqs. (5.7–5.11) are assumed to be images represented in matrix form and introduced inside the CFPI, where $P(\rho) = \phi(x, y) = \phi(m\Delta x, n\Delta y)$. The image matrix has dimensions of $M \times N$, where $m = 1: 512$ and $n = 1: 512$ pixels. Then, the fringe shift present in the image is in the harmonic term in the Airy distribution. Hence, Eq. (5.2) can be rewritten as follows:

$$I(\delta; R, N) = [\frac{T^2}{1 + R^2 - 2R\cos[\delta + \varphi(x, y)]}]^N \qquad (5.12)$$

By using equation (5.4), we can write the normalized intensity considering the deformed phase as follows:

$$I_{normalized}(\delta; R, N) = \frac{I}{I_{max}} = \frac{1}{[1 + F\,sin^2(\frac{\delta+\varphi(x,y)}{2})]^N} \qquad (5.13)$$

where F is defined in equation (5.5).

Now, we compute the phase mapping of the image from Eq. (5.12) as follows:

The deformed or shifted phase is the sum of the interfringe spacing and the image phase represented as follows:

$$\varphi(x, y)_{deformed} = \delta + \varphi(x, y) \qquad (5.14)$$

Consequently, the phase of the image is easily deduced from Eq. (5.13) as follows:

$$\varphi(x, y) = \varphi(x, y)_{deformed} - \pi \qquad (5.15)$$

where the interfringe spacing is given by:

$$\delta = \frac{2\pi}{\lambda}\,(\text{O.P.D.}) = \frac{2\pi}{\lambda}\left(\frac{\lambda}{2}\right) = \pi. \qquad (5.16)$$

Compared with the known method of fringe shift obtained at 0, $\pi/2$, π, and $3\pi/2$ for successive intensities I_1, I_2, I_3, and I_4, this result is given as follows:

$$\varphi(x, y) = \tan^{-1}\left(\frac{I_2 - I_1}{I_4 - I_3}\right) \qquad (5.17)$$

5.3   Results and Discussion

The normalized coefficient of finesse F as a function of reflectivity R for different cascaded interferometers is computed from Eq. (5.5) and represented as shown in Fig. 5.2. $N = 1$ represents the ordinary FPI, while $N = 2, 3$, and 4 represent the number of cascaded interferometers.

The investigated linear, conical, quadratic, and Gaussian apertures are designed using MATLAB code. The linearly distributed aperture with a diameter of 256 pixels and the matrix dimensions of the whole image 512×512 pixels are shown in Fig. 5.3.

Fig. 5.2 The normalized coefficient of finesse F as a function of reflectivity R for different cascaded interferometers. $N = 1$ represents the ordinary FPI, while $N = 2$, 3, and 4 represent the number of cascaded interferometers

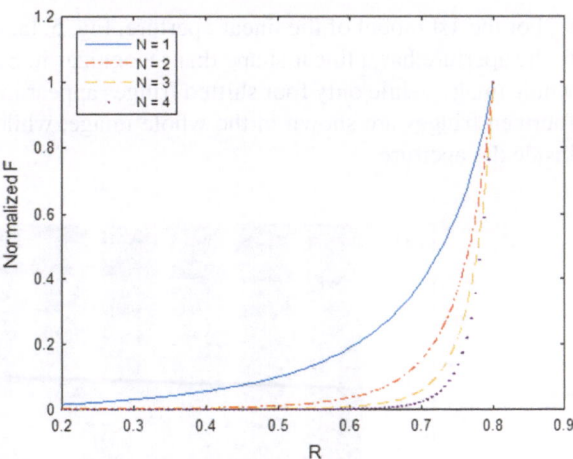

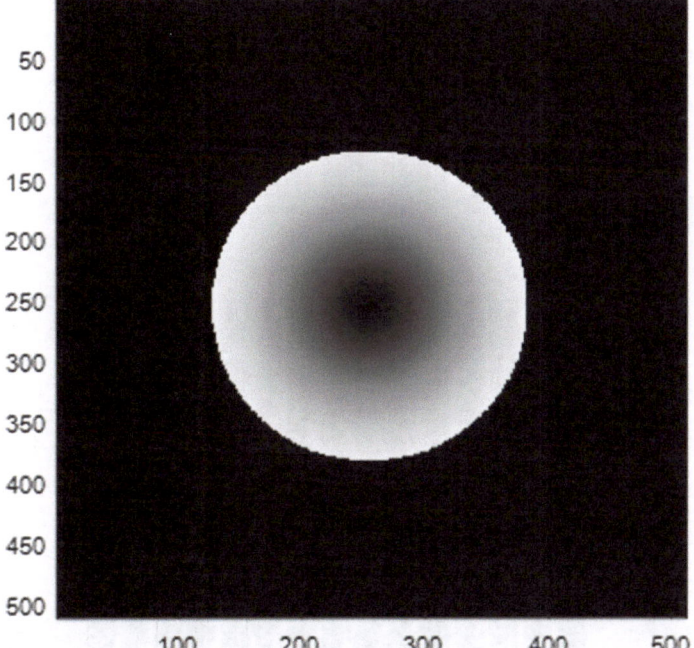

Fig. 5.3 Linearly distributed aperture with a diameter of 256 pixels, the matrix dimensions of the whole image are 512×512 pixels

For the 1st model of the linear aperture, Fig. 5.4a, shows that the fringe shift due to the aperture has a linear shape that recognizes it. Seven fringes are shown for the whole image, while only four shifted fringes appear inside the aperture. In Fig. 5.4b, fourteen fringes are shown in the whole image, while seven shifted fringes appear inside the aperture.

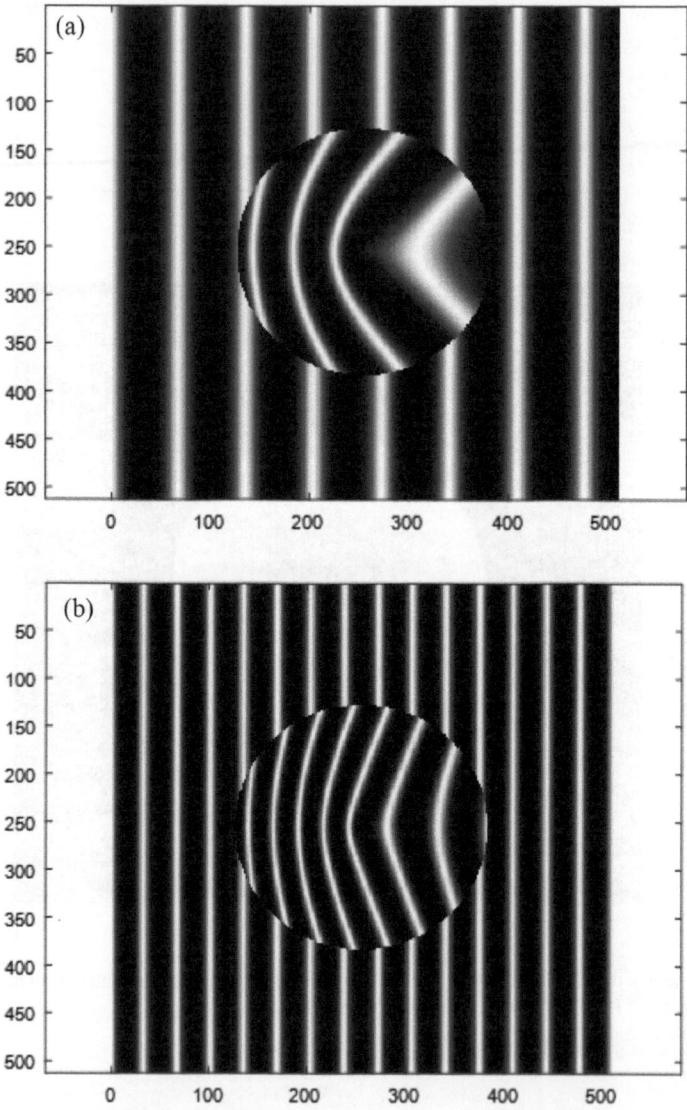

Fig. 5.4 **a** Fringe shift has a linear shape to identify the linear aperture. Seven fringes are shown for the whole image, while only four fringes appear inside the aperture. **b** Fourteen fringes are shown in the whole image, while seven shifted fringes appear inside the aperture

In one dimension, the fringe shift corresponding to the linear aperture is shown in Fig. 5.5a. The fringes give the phase of the image at different locations, whereas the 2nd fringe, from the right, gives zero shift. Seven fringes are shown in the plot, while only four shifted fringes appear inside the aperture. For the linear aperture, the fringe shift equals the interfringe spacing ($\delta Z = \Delta Z$). In Fig. 5.5b, the 3rd fringe has a zero-shift compared with the other straight-line fringes, giving the phase of the image at different locations. Fourteen fringes are shown in the plot, while only eight shifted fringes appear inside the aperture.

For the 2nd model, namely the conic aperture where $P(\rho) = 1 - \rho$, the fringe shift has a conic shape, as shown in Fig. 5.6. The shift appears in the reverse direction corresponding to the linear aperture. The aperture has a diameter $= 256$ pixels, and the matrix dimensions of the whole image are 512×512 pixels. Seven fringes are shown for the whole image, while only four fringes shifted to the right appear. The fringe shift in one dimension has a conical shape, as shown in Fig. 5.7. The shift appears in the reverse direction corresponding to the linear aperture. The aperture has a diameter $= 256$ pixels, and the matrix dimensions of the whole image are 512×512 pixels. Four shifted fringes are shown inside the conic aperture. The MATLAB code for the conic aperture is written as follows:

The code for the monodimensional conic aperture is written as follows:

$$A(r, c) = 1 - (r - x_{\text{cen}})$$

An aperture in the form $P(\rho) = \rho^{1.1}$ is shown in Fig. 5.8a. This aperture deviates from linearity by 10%. The corresponding fringe shift is not linear, hence, any deviation from linearity results in a quadratic fringe shift, as shown in Fig. 5.8b.

The MATLAB code for the aperture deviated from the linear distribution by 10%:

$$A(r, c) = \left(\text{sqrt}\left((r - x_{cen})\text{^}2 + (c - y_{cen})\text{^}2\right)\right)\text{^}1.1$$

The investigation of the fringe shift for this aperture in one dimension is outlined in Fig. 5.9a–c. The fringe shift in the case of one direction for a nearly linear aperture is shown in Fig. 5.9a, where the cursor stands at point $x_1 = 272$ pixels for the 4th fringe from the right. In Fig. 5.9b, the cursor stands at the point $x_1 = 204$ pixels for the 5th fringe from the right. Hence, the interfringe spacing is deduced to be $\Delta Z = 272 - 204 = 68$ pixels. Figure 5.9c represents the fringe shift in the case of one direction for a nearly linear aperture, and the cursor stands at the point $x_1 = 197$ pixels for the shifted fringe. The amount of fringe shift for this deviated aperture from linearity is: $\delta Z_d = 272 - 197 = 75$ pixels. The differential fringe shift is: $(\delta Z_d - \Delta Z)/\Delta Z = (75 - 68)/68 = 0.102$.

The amount of fringe shift for this deviated aperture from linearity is: $\delta Z_d = 272 - 197 = 75$ pixels. The differential fringe shift is:

$(\delta Z_d - \Delta Z)/\Delta Z = (75 - 68)/68 = 0.102$. This differential fringe shift is related to the deviation of the aperture from linearity.

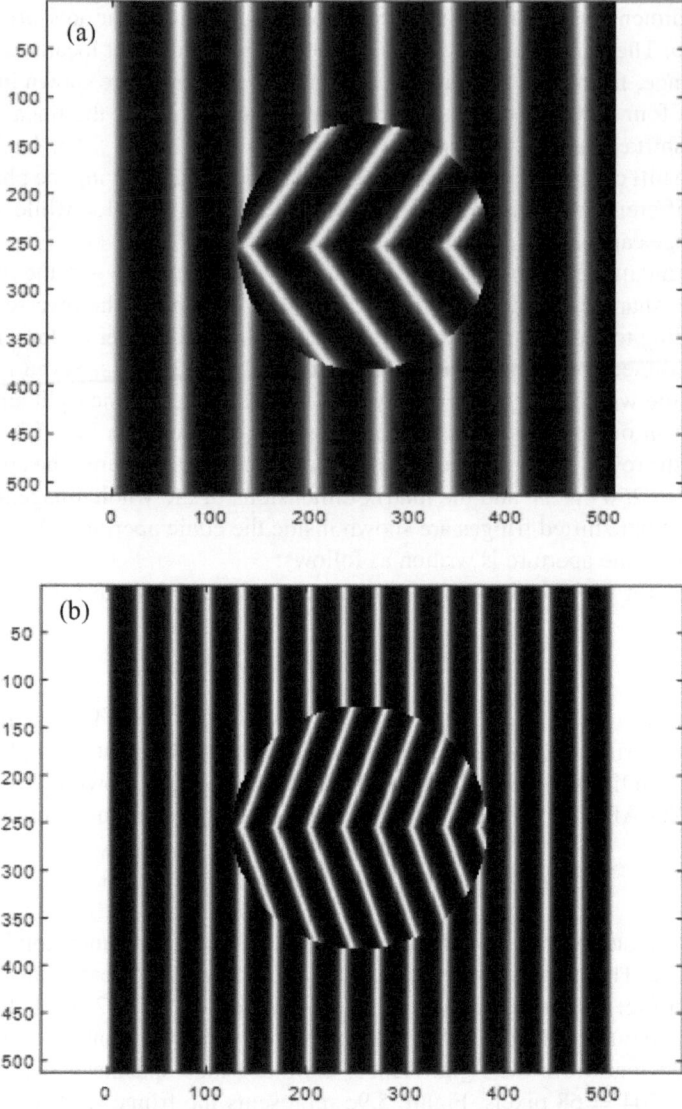

Fig. 5.5 a Fringe shift corresponding to the linear aperture in one dimension. The fringes give the phase of the image at different locations, while the 2nd fringe gives zero shift. Seven fringes are shown in the plot, while only four fringes appear inside the aperture. For the linear aperture, the fringe shift equals the interfringe spacing ($\delta Z = \Delta Z$). **b** Fringe shift in one dimension is shown in the graph. The 3rd fringe has a zero-shift compared with the other straight-line fringes, giving the phase of the image at different locations. Fourteen fringes are shown in the plot, while only eight fringes appear inside the aperture

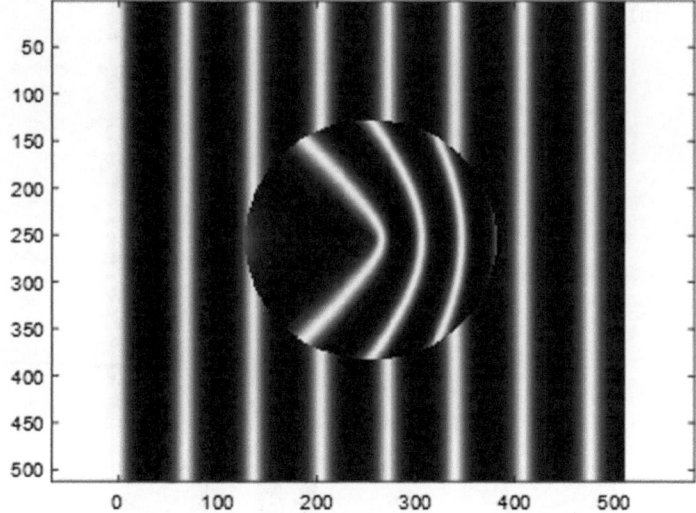

Fig. 5.6 Fringe shift has a conical shape. The shift appears in the reverse direction corresponding to the linear aperture. The aperture has a diameter = 256 pixels, and the matrix dimensions of the whole image are 512 × 512 pixels. Seven fringes are shown

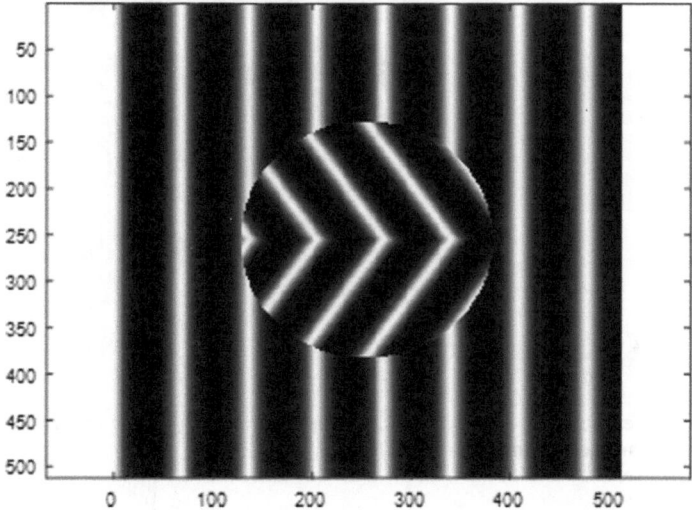

Fig. 5.7 Fringe shift in one dimension has a conical shape. The shift appears in the reverse direction corresponding to the linear aperture. The aperture has a diameter = 256 pixels, and the matrix dimensions of the whole image are 512 × 512 pixels. Four fringes are shown inside the conic aperture

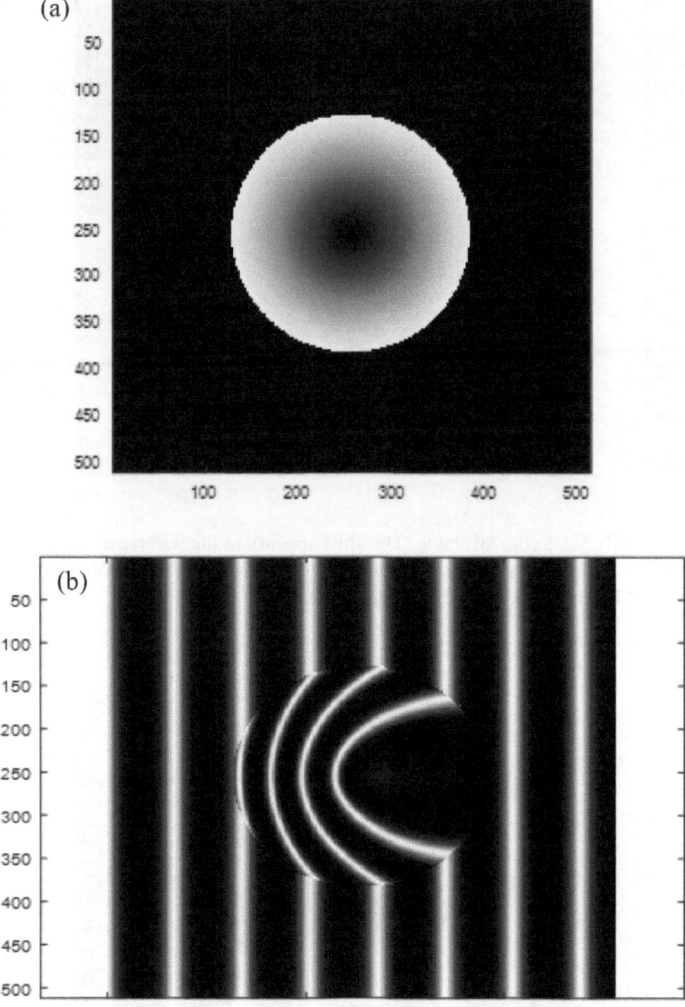

Fig. 5.8 **a** An aperture in the form $P(\rho) = \rho^{1.1}$ is shown in the graph. **b** Quadratic fringe shift corresponding to an aperture in the form $P(\rho) = \rho^{1.1}$

For the 3rd model, quadratic aperture with a diameter of 256 pixels was used is shown in Fig. 5.10a. Interference fringes modulated by the monodimensional quadratic aperture with a diameter of 256 pixels are shown in Fig. 5.10b.

The MATLAB code of the aperture is represented as follows:

$A(i,j) = \frac{(i-x_{\text{cen}})^2}{256}$, where $i = 1{:}512, j = 1{:}512$ pixels, and $x_{\text{cen}} = 256$ pixels.

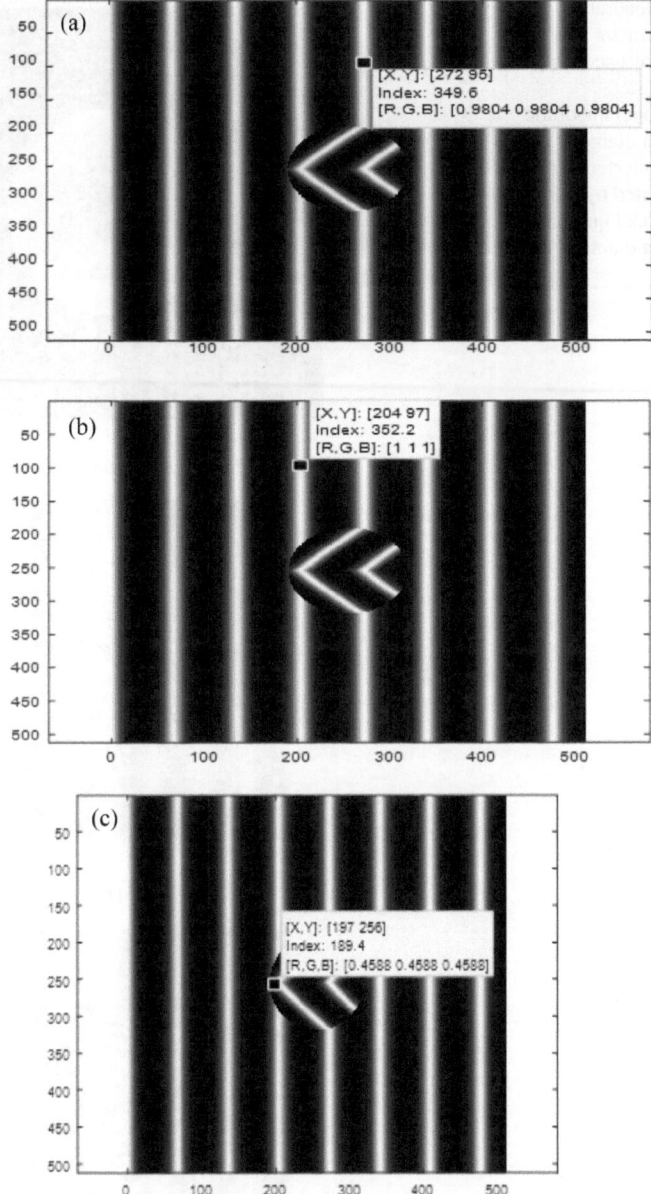

Fig. 5.9 a Fringe shifts in case one direction for a nearly linear aperture, and the cursor stands at point $x_1 = 272$ pixels for the 4th fringe from the right. **b** Fringe shifts in case one direction for a nearly linear aperture, and the cursor stands at point $x_1 = 204$ pixels for the 5th fringe from the right. The interfringe spacing is deduced from Fig. 5.7a,b as: $\Delta Z = 272–204 = 68$ pixels. **c** Fringe shifts in the case of one direction for a nearly linear aperture, and the cursor stands at point $x_1 = 197$ pixels for the shifted fringe

Fig. 5.10 **a** Quadratic aperture of diameter 256 pixels. **b** Interference fringes modulated by a monodimensional quadratic aperture with a diameter of 256 pixels. **c** Interference fringes modulated by the monodimensional quadratic aperture with a diameter of 256 pixels

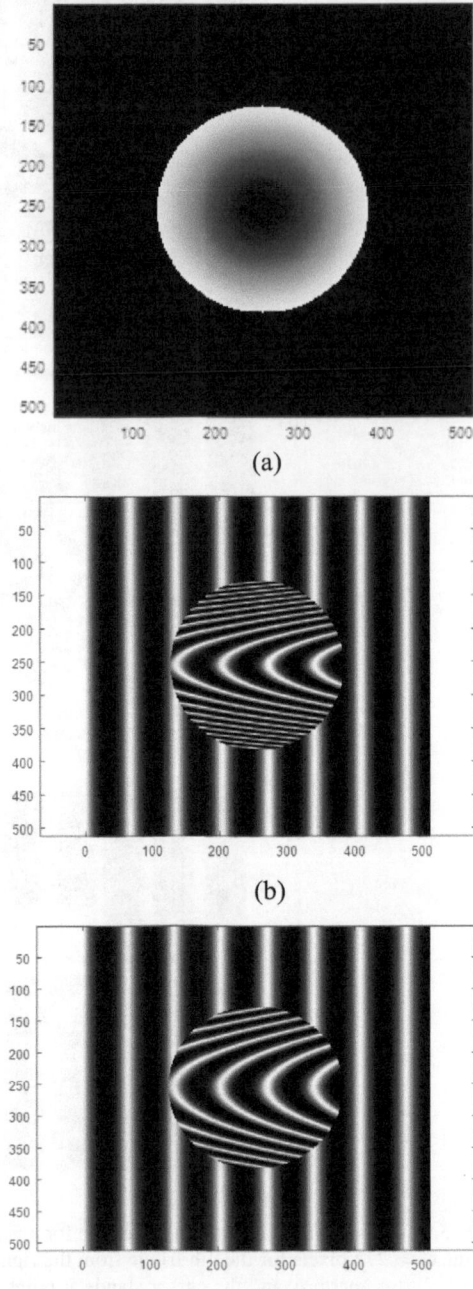

Another interference fringe modulated by the monodimensional quadratic aperture with a diameter of 256 pixels of different amplifications is shown in Fig. 5.10c. The corresponding MATLAB code of the aperture is:

$A(i, j) = (i - x_{cen})^2/1024$, where i, j, and x_{cen} are shown in Fig. 5.10b.

The 4th model of the Gaussian aperture of truncation radius $R = 64$ pixels is shown in Fig. 5.11a. The MATLAB code for the Gaussian aperture is written as follows:

$$A(i, j) = G * \exp(-((i - x_{cen})^\wedge 2 + (j - y_{cen})^\wedge 2)/R^\wedge 2); \ G = 4$$

The fringe shift corresponding to the Gaussian aperture inside a circle with a diameter of 256 pixels is shown in Fig. 5.11b. The aperture has a radius $R = 64$ pixels and $G = 4$. The fringe shift for the Gaussian aperture for $G = 8$, is shown in Fig. 5.11c. Another plot of the fringe shift inside a circle with a diameter of 512 pixels for a Gaussian aperture of radius $R = 128$ pixels is shown in Fig. 5.11d. It is shown from all the plots shown in Fig. 5.11b, c that the distribution is Gaussian.

The MATLAB code for the Gaussian aperture

$$A(i, j) = (4) * \exp(-((i - x_{cen})^\wedge 2 + (j - y_{cen})^\wedge 2)/R^\wedge 2);$$

$$A(i, j) = G * \exp(-((i - x_{cen})^\wedge 2 + (j - y_{cen})^\wedge 2)/R^\wedge 2); \ G = 4$$

The code for the fringe shift corresponding to the Gaussian aperture is:

$$F1 = 4 * r1^\wedge 2 * (\sin(0.5 * y - A(i, j))^\wedge 2);$$

The B/W concentric annuli are shown in Fig. 5.12a. It has four equally transparent annuli in succession with four equally black annuli. The fringe shift corresponding to the B/W concentric annulus shifted discontinuous straight-line fringes, as shown in Fig. 5.12b.

The comparison of the fringe shift for the modulated apertures with the uniform circular aperture is shown in Fig. 5.13a,b. A uniform circular aperture of constant transmittance with a diameter of 256 pixels is shown in Fig. 5.13a, while a shifted straight-line fringe of constant value identifies the uniform circular aperture, as shown in Fig. 5.13b.

5.4 Conclusion

We conclude that the selected aperture models are recognized by referring to the corresponding fringe shift. A linear fringe shift corresponding to the linearly distributed aperture is shown. The aperture with a conic distribution has a similar

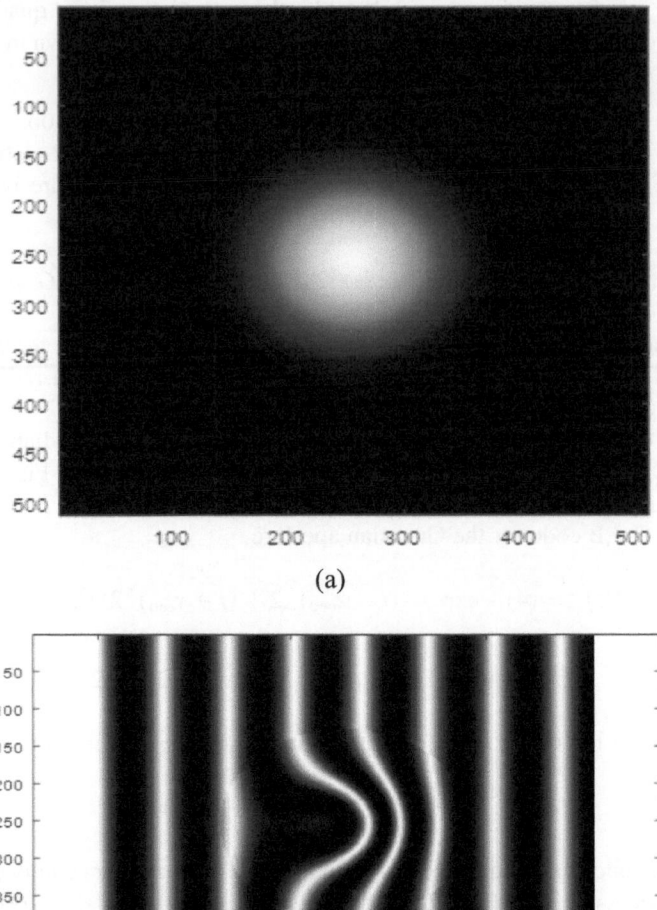

(a)

(b)

Fig. 5.11 **a** Gaussian aperture of truncation radius $R = 64$ pixels. **b** Fringe shift corresponding to the Gaussian aperture inside a circle with a diameter of 256 pixels. The aperture radius is $R = 64$ pixels. **c** $A(i, j) = G * \exp(-((i - x_{cen})^2 + (j - y_{cen})^2)/R^2); \ G = 8.$ **d** Fringe shift inside a circle with a diameter of 512 pixels for a Gaussian aperture of radius $R = 128$ pixels

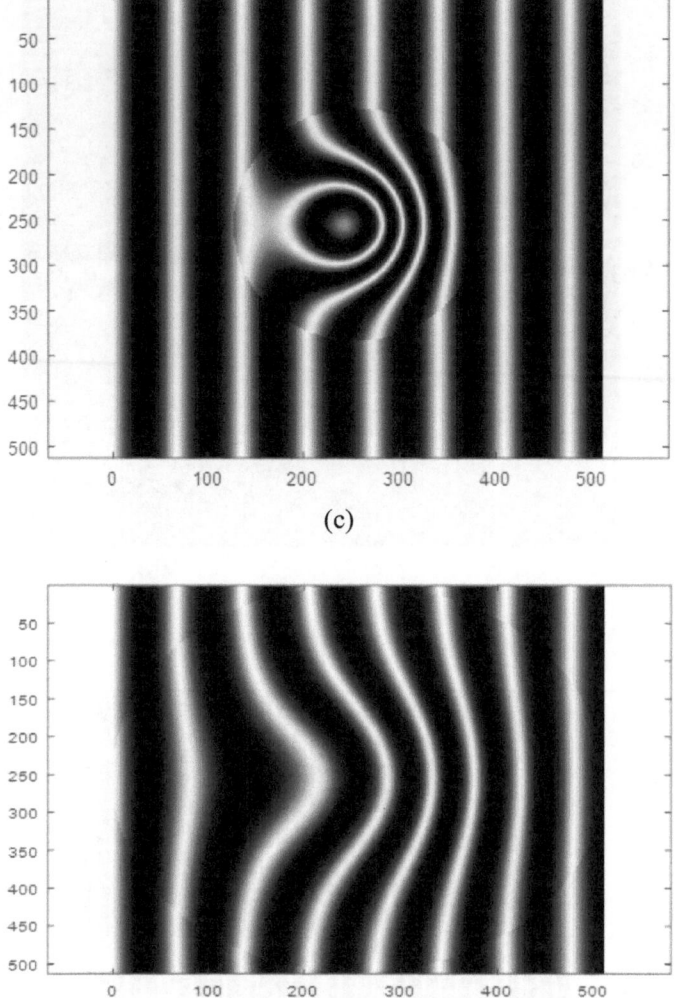

(c)

(d)

Fig. 5.11 (continued)

linear fringe shift corresponding to the linearly distributed aperture but in the opposite direction. The quadratic fringe shift manifests the recognition of quadratically distributed apertures. Finally, the Gaussian fringe shift recognizes the Gaussian aperture, which is characteristic of laser beam propagation. The B/W concentric annuli have constant discontinuous fringe shifts. A comparison with a uniform circular aperture, which results in a constant shift, is given. A deviation from the linearly distributed aperture of 10% transforms the fringe shift from linear to quadratic variation.

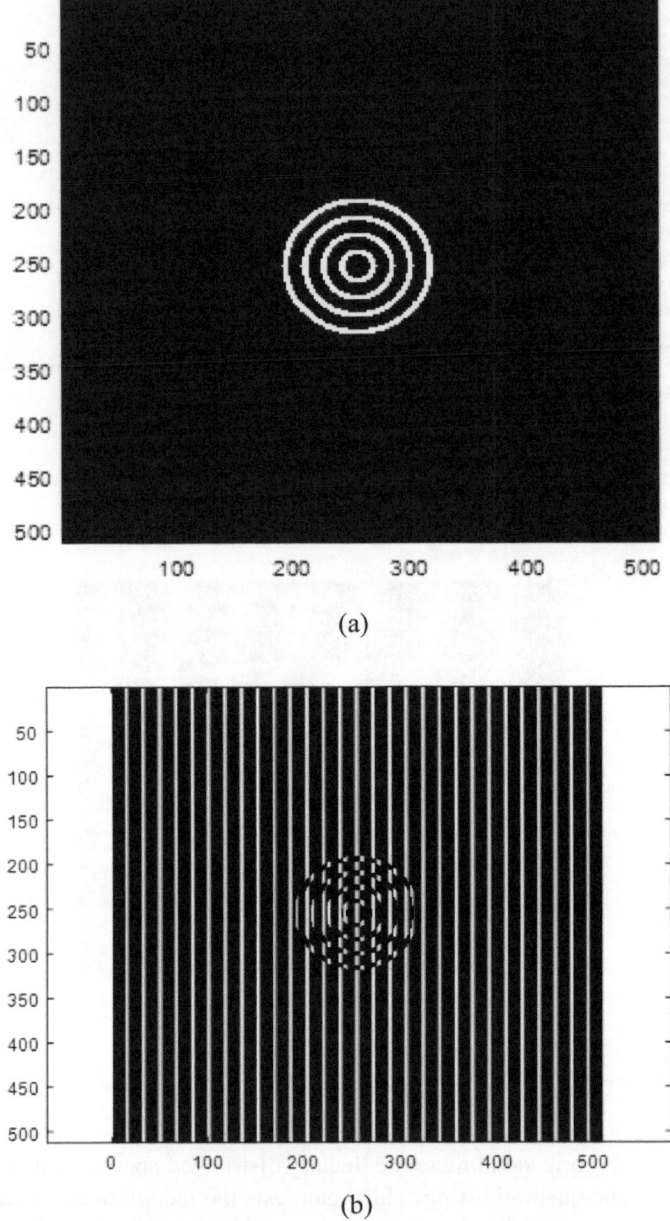

(a)

(b)

Fig. 5.12 **a** Aperture of the B/W concentric annulus. It has four equally transparent annuli in succession with four equally transparent black annuli. **b** Fringe shift corresponding to B/W concentric annuli. It has shifted straight-line fringes

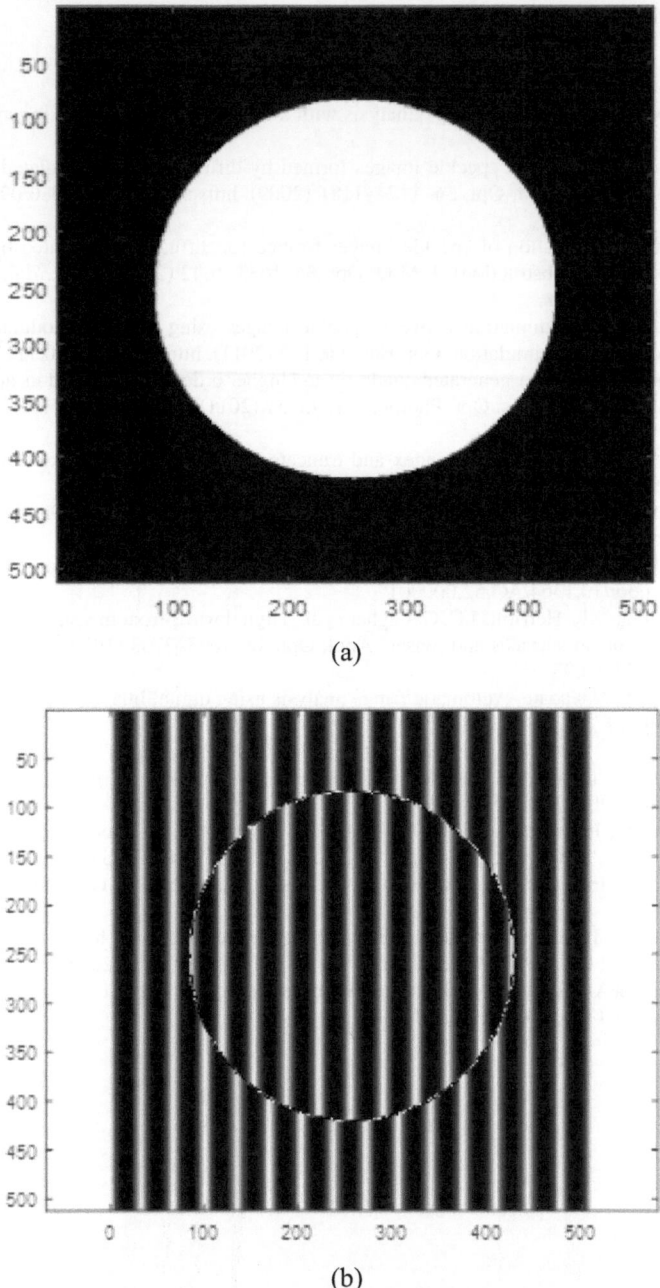

(a)

(b)

Fig. 5.13 **a** Uniform circular aperture with constant transmittance. The aperture has a diameter of 256 pixels. **b** Shifted straight-line fringes of constant value identify the uniform circular aperture shown in (**a**)

References

1. K. Chandra Shakher, K. Matsuda et al., *Proceedings of ICICS* (1997)
2. D.W. Robinson, Automatic fringe analysis with a computer image processing system. Appl. Opt. **22**, 2169–2176 (1983)
3. A.M. Hamed, Numerical speckle images formed by diffusers using modulated conical and linear apertures. J. Mod. Opt. **56**, 1174–1181 (2009). https://doi.org/10.1080/09500340902985379
4. A.M. Hamed, Formation of speckle images formed for diffusers illuminated by modulated apertures (circular obstruction). J. Mod. Opt. **56**, 1633–1642 (2009). https://doi.org/10.1080/09500340903277792
5. A.M. Hamed, Discrimination between speckle images using diffusers modulated by some deformed apertures: simulation. Opt. Eng. **50**, 1–7 (2011). https://doi.org/10.1117/1.3530085
6. A.M. Hamed, Computer generated quadratic and higher order apertures and its application on numerical speckle images. Opt. Photon. J. **1**, 43–51 (2011). https://doi.org/10.4236/opj.2011.12007
7. A.M. Hamed, Study of graded index and truncated apertures using speckle images. Precis. Instr. Mechanol. (PIM) **3**, 144–152 (2014)
8. A.M. Hamed, Image processing of corona virus using interferometry. Opt. Photon. J. **6** (2016). https://doi.org/10.4236/opj.2016.65011
9. J.C. Wyant, Computerized interferometric surface measurements. Appl. Opt. **52**, 1–8 (2013). https://doi.org/10.1364/AO.52.000001
10. J.H. Bruning, D.R. Herriott, J.E. Gallagher et al., Digital wave-front measuring interferometer for testing optical surfaces and lenses. Appl. Opt. **13**, 2693–2703 (1974). https://doi.org/10.1364/AO.13.002693
11. T. Yatagai, S. Nakadate, Automatic fringe analysis using digital image processing technique. Optic. Eng. **21**, 432–435 (1982)
12. A.M. Hamed, www.lap.com (Lambert Academic Publishing), The point spread function of some modulated apertures (Application on speckle and interferometry images) ISBN:9786202070706 (2017)
13. A.M. Hamed, Investigation of SIDA Virus (HIV) images using interferometry and speckle techniques. Int. J. Innov. Res. Comput. Sci. Tech. (IJIRCST) **4**, 38–45 (2016)
14. A.M. Hamed, Image processing of coronavirus using interferometry. Opt. Photon. J. **6**, 75–86 (2016)
15. A.M. Hamed, Topics on optical and digital image processing using holography and speckle techniques, publisher by Lulu.com, ISBN 9781329328464 Nov. 29 (2015)
16. A.M. Hamed, A modified michelson interferometer and an application on microscopic imaging. Int. J. Photon. Optic. Technol. (IJPOT) **3** (2017)

Part II
Application of Medical Images Using Cascaded Interferometers

Chapter 6
Processing of Retinal Artery Images Using Higher Orders of Two-Beam Interference

Coherent illumination is assumed in the fabrication of interferometer fringes modulated by the considered image. The modulated fringes resulted from the multiplication of the ordinary fringe system and the retinal arterial image. Digital cascaded two-beam interference using feedback rays is considered to give a cosine function of higher order n greater than one ($n > 1$).

The processing of retinal artery images of normal and occluded arteries was investigated using digital two-beam interference of higher orders. Five segments from the retinal arteries are considered. The higher order power of two-beam interference is compared with that of ordinary two and multiple beam interference. The refractive index distribution of the retinal artery is extracted from the fringe shift that occurs in the artery.

All images of the retina used in this chapter were taken from the following site [1]:

(https://webvision.med.utah.edu/book/part-i-foundations/simple-anatomy-of-the-retina/).

6.1 Introduction

Retinal image analysis, concepts, and applications are presented in a review article [2]. Image enhancement, segmentation, and restoration are the main topics of this article. In addition, a dimensionless measure of retinal photography is presented. Morphological image processing exploits features of the vasculature shape that are known a priori, such as being piecewise linear and connected. In this work, and others [3–6] retinal vessel widths were measured. The diameter of the central artery is computed from noninvasive measurements in humans [4]. An accurate assessment of changes in retinal vessel diameter using multiple-frame electrocardiograph synchronized fundus photography is outlined in [5]. In addition, retinal blood vessel

© The Author(s), under exclusive license to Springer Nature Switzerland AG 2024   65
A. Hamed, *Cascaded Interferometers and Their Medical Applications*,
SpringerBriefs in Applied Sciences and Technology,
https://doi.org/10.1007/978-3-031-64535-8_6

width was measured using computerized image analysis [6]. Another work on the quantification and characterization of arteries in retinal images is presented in [7]. Reproducible estimation of retinal vessel width by computerized microdensitometry was described in [8].

Retinal artery occlusion can occur in either the central retinal artery or in a branch retinal artery that splits off the central retinal artery. Either the artery can become blocked by a clot or "embolus" in the bloodstream. Retinal artery occlusion is considered a medical emergency and requires immediate attention. When artery occlusion occurs, the oxygen supply to the area of the retina nourished by the affected artery decreases, causing permanent vision loss. Most patients who suffer retinal artery occlusion are between the ages of 50 and 80 years. They notice a sudden, painless loss of vision that can be a complete loss of vision if it is a central retinal artery occlusion or a partial loss of their visual field if it is a branch retinal artery occlusion [9]. In general, retinal arteries may become blocked when a blood clot or fat deposits become stuck in the arteries. These blockages are more likely if there is hardening of the arteries (atherosclerosis) in the eye. Clots may travel from other parts of the body and block an artery in the retina. The most common sources of clots are the heart and carotid artery in the neck [10, 11].

The principle of applying OCT to retinal images is based on interferometry [12–15]. In a typical early-generation OCT system, visible light (i.e., to visualize the beam) and broadband, short-coherence length, near-IR light are coupled into one branch of a Michelson interferometer. The light is then split into two paths, one leading to a reference mirror, and the other focused onto the retina. Light is reflected and backscattered from refractive index interfaces within the retina according to the optical properties of each interface. The reflected light from the retina (i.e., the sample arm) and from the reference mirror is recoupled into the interferometer to ultimately be detected after interference in the spectrometer. Using "time domain" OCT, reflection sites at various depths in the tissue can be sampled by changing the path length of the reference arm.

In this chapter, image processing of retinal arteries is realized by interferometry. The method is based on using microscopy images of the retina and placing them in the interference cosine term. Higher-order two-beam interference is suggested by the author to obtain sharper fringes than ordinary two and multiple beam interference. We computed the refractive index of different segments from the arteries considering the fringe shift that occurred to interfringe spacing [16] in the modulated trigonometric function.

6.2 Theoretical Analysis

Before we started with interferometer processing using normal and occluded retinal images, we obtained cross sections of retinal images, as shown in Fig. 6.1. The retina has arteries and veins. In addition, the optic disc is shown. Higher-order two-beam interference is described using feedback rays, as shown in Fig. 6.2. The interference

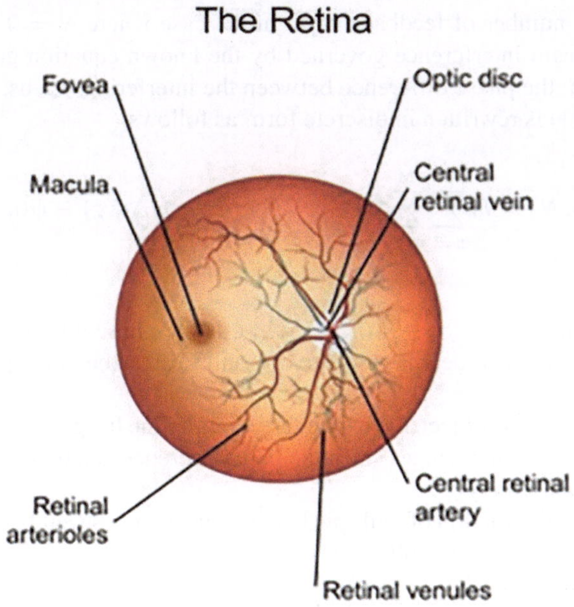

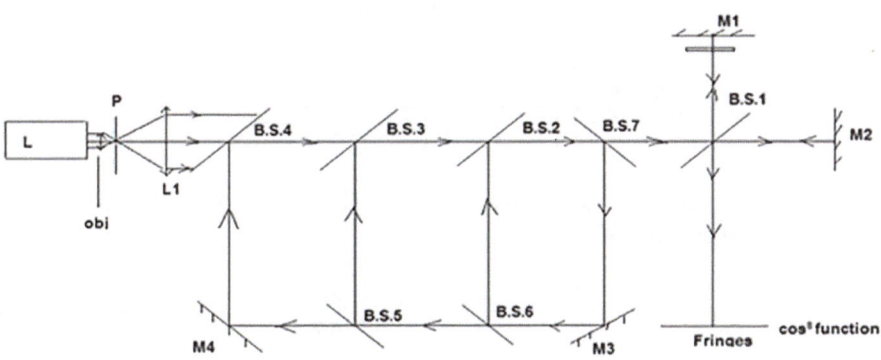

Fig. 6.1 Image of the retina obtained through an ophthalmoscope

Fig. 6.2 Higher-order two-beam interference considering four feedback rays giving a $\cos^8$ function instead of a $\cos^2$ function. B. S_1, B. S_2, B. S_3, B. S_4, B. S_5, B. S_6, and B. S_7 are beam splitters, while M_1, M_2, M_3, and M_4. L: coherent laser beam, P pinhole, and L_1 collimating lens placed in the focal plane to the pinhole to render the beam parallel

pattern is governed by the following equation:

$$I_{\text{feed back}}(x, y; N) = I_0 \cos^{2(N+1)}(\delta) \tag{6.1}$$

where N is the number of feedback rays. In the case where $N = 0$, we obtain the ordinary two-beam interference governed by the known equation governed by the $\cos^2$ function. δ: the phase difference between the interfering beams.

Equation (6.1) is rewritten in discrete form as follows:

$$I_{\text{feed back}}(x, y, N) = I_0 \sum_{n=1}^{N} \sum_{m=1}^{M} \{\cos^{2(N+1)}(\Phi(n\Delta x, m\Delta y; z) - \psi(n\Delta x, m\Delta y))\}$$

(6.2)

where $\delta = \Phi(n\Delta x, m\Delta y; z) - \Psi(n\Delta x, m\Delta y)$, is the phase difference between the object and the reference beams respectively, and O.P.D. the corresponding optical path difference.

In the absence of the object, the fringes are straight line fringes, while the presence of the object deforms the fringes since the path difference is a function of the object phase information $\Phi(x, y)$.

The intensity distribution of multiple beam interference is well known as the Airy pattern and is represented as follows [16]:

$$I(x, y, z) = I_0 \frac{1}{1 + F \sin^2\left(\frac{\delta}{2}\right)}; \delta = \left(\frac{2\pi}{\lambda}\right) \text{O.P.D}$$

(6.3)

Equation 6.3 is rewritten in discrete form as follows:

$$I(x, y, z) = I_0 \sum_{n=1}^{N} \sum_{m=1}^{M} \left\{ \frac{1}{\left[1 + F \sin^2\left(\frac{\Phi(n\Delta x, m\Delta y; z) - \psi(n\Delta x, m\Delta y)}{2}\right)\right]} \right\}$$

(6.4)

The refractive index of the microscopy image is computed from the interference fringe shift [16] as follows:

$$\mu(x, y) = 1 + a(x, y)\frac{\delta Z}{\Delta Z}$$

(6.5)

δZ is the fringe shift, Δz is the interfringe spacing, and $a(x, y)$ represents the amplitude of the image that lies between 0 and 1.

The mean refractive index and the S.D. or the root mean square value are computed using these known formulae:

$$\langle \mu \rangle = \frac{1}{N} \sum_{i=1}^{N} \mu_i(x, y) \quad , \sigma = \sqrt{\sum_{i=1}^{N} [\mu_i(x, y) - \langle \mu \rangle]^2}$$

(6.6)

6.3 Results and Discussion

Retinal images with numbers placed beside the investigated arteries in the retina are shown in Fig. 6.3. Five segments from the retinal arteries are shown in Fig. 6.4. All the images were resized to have dimensions of 512×512 pixels.

Modulated ordinary two-beam interference of retinal artery 1 is computed using Eq. 6.2, where $N = 0$, i.e., no feedback rays. The fringes are plotted in Fig. 6.5a. The trajectory of the artery is recognized from the shift that occurred in the image, and the refractive index of the retinal artery is computed from the fringe shift using Eq. 6.5. The modulated multiple beam interference of the retinal artery [2] is computed using Eq. (6.4) in matrix form and plotted as shown in Fig. 6.5b. The fringes are sharper than the two-beam fringes, as expected. Now, the two-beam interference of feedback rays considering $N = 25$ is computed from Eq. 6.2 which gives sharper fringes than multiple beam fringes governed by the Airy distribution Eq. 6.4.

Similar plots are given for the other arteries outlined by 2, 3, 4, and 5 and plotted for the interference of the two and multiple beams compared with the sharper fringes given by two beams of higher orders N. These plots are shown in the Figs. 6.6, 6.7, 6.8, 6.9. Again, the trajectory of the arteries recognized from the shift occurred in the interference image.

Fig. 6.3 Retinal image with numbers placed beside the investigated arteries of dimensions 512×512 pixels corresponding to 2×2 cm

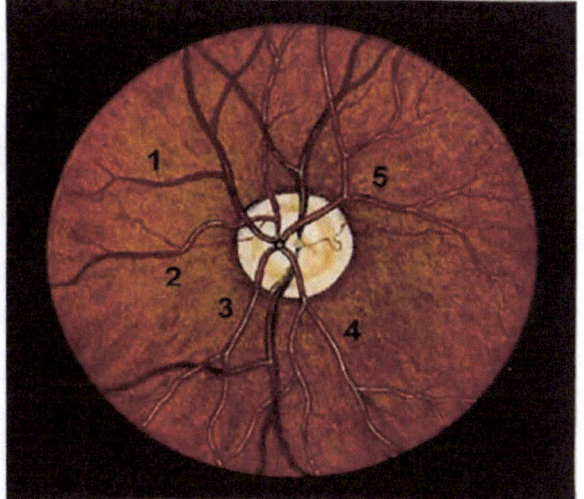

Fig. 6.4 From above, five segments from retinal arteries 1, 2, 3, 4, and 5 with resized dimensions of 512 × 512 pixels are shown

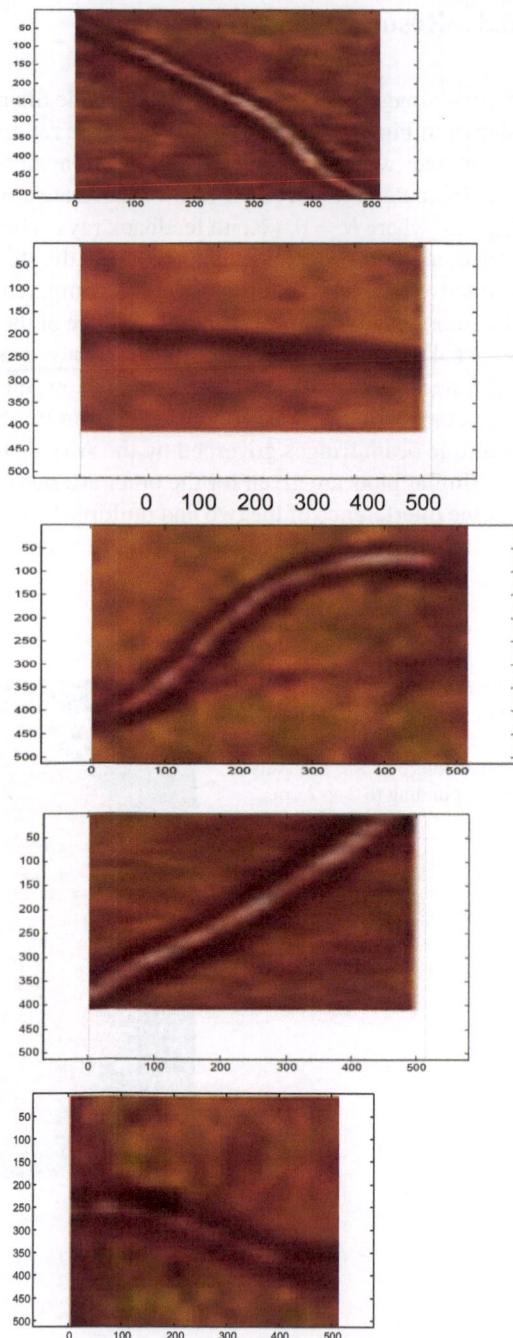

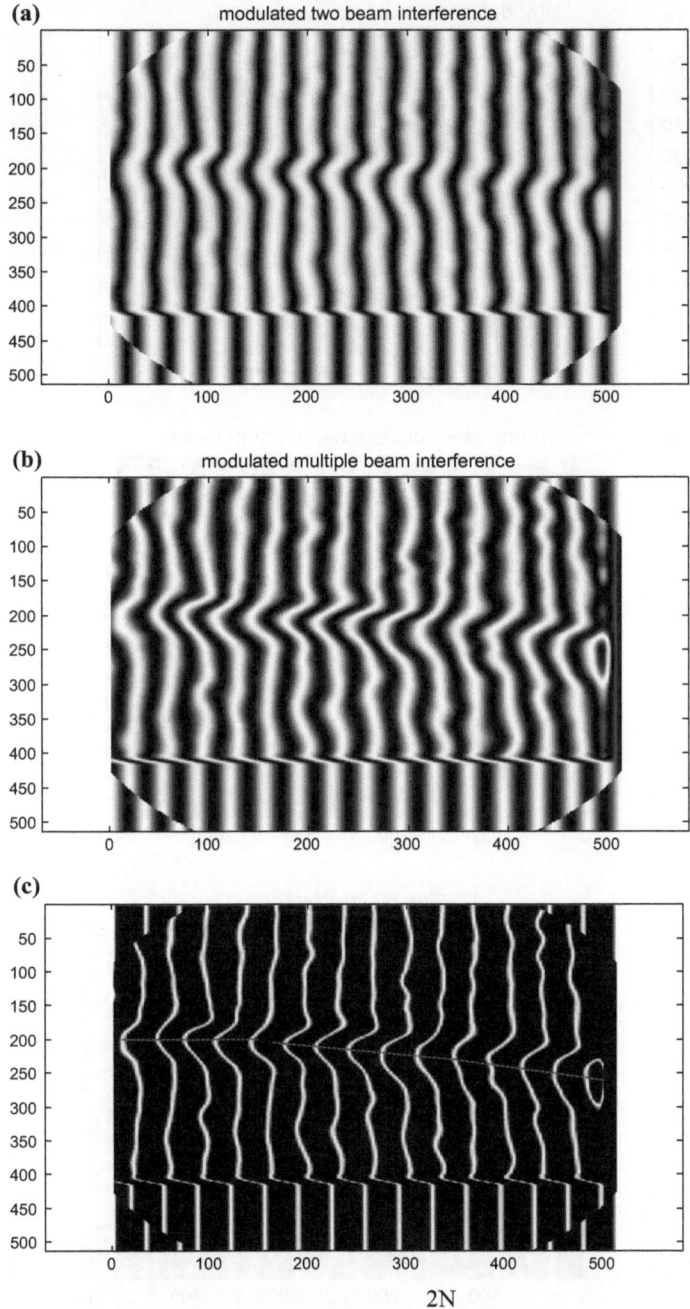

Fig. 6.5 **a** Modulated ordinary two-beam interference of retinal artery 1. **b** Modulated multiple beam interference of retinal artery 1. **c** Two-beam interference of higher order $\cos^{2N}$, where $N = 25$. The discontinuous red line is placed on the center of the fringe shift of the segment from artery 1

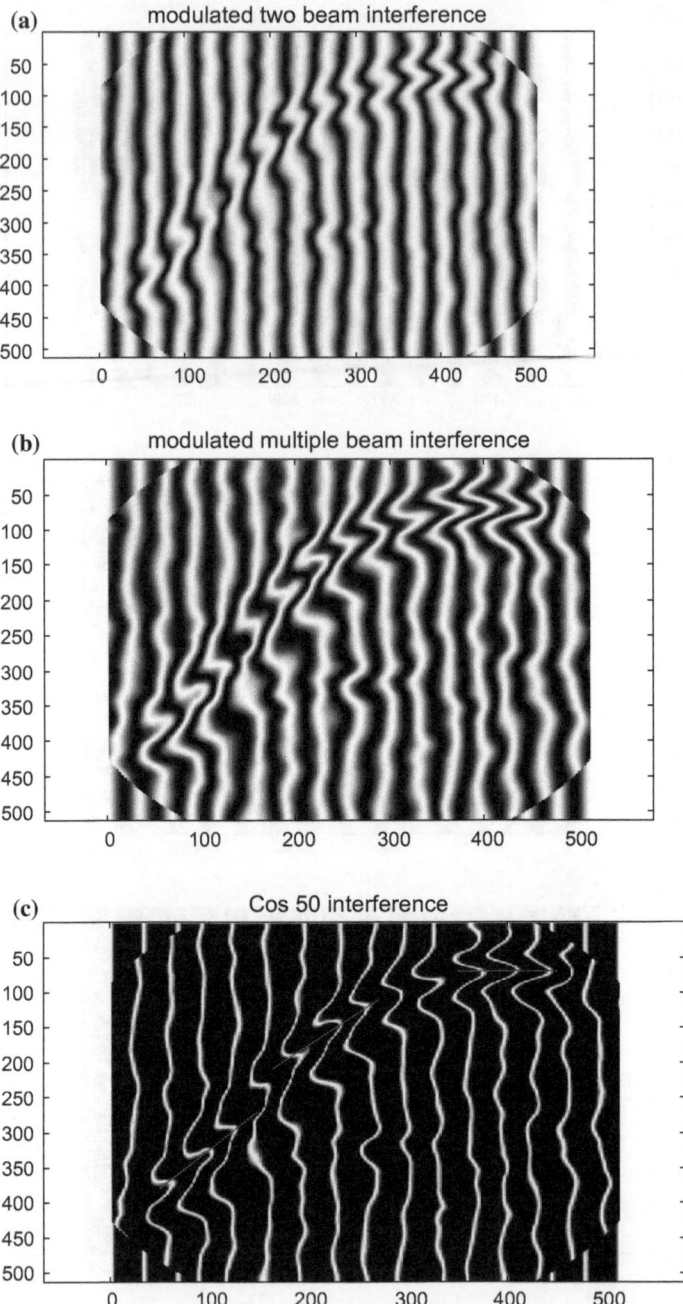

Fig. 6.6 **a** Modulated ordinary two-beam interference of retinal artery 2. **b** Modulated multiple beam interference of retinal artery 2. **c** Two-beam interference of higher power $\cos^{2N}$, where $N = 25$. The discontinuous red line is placed on the center of the fringe shift of the segment from the artery

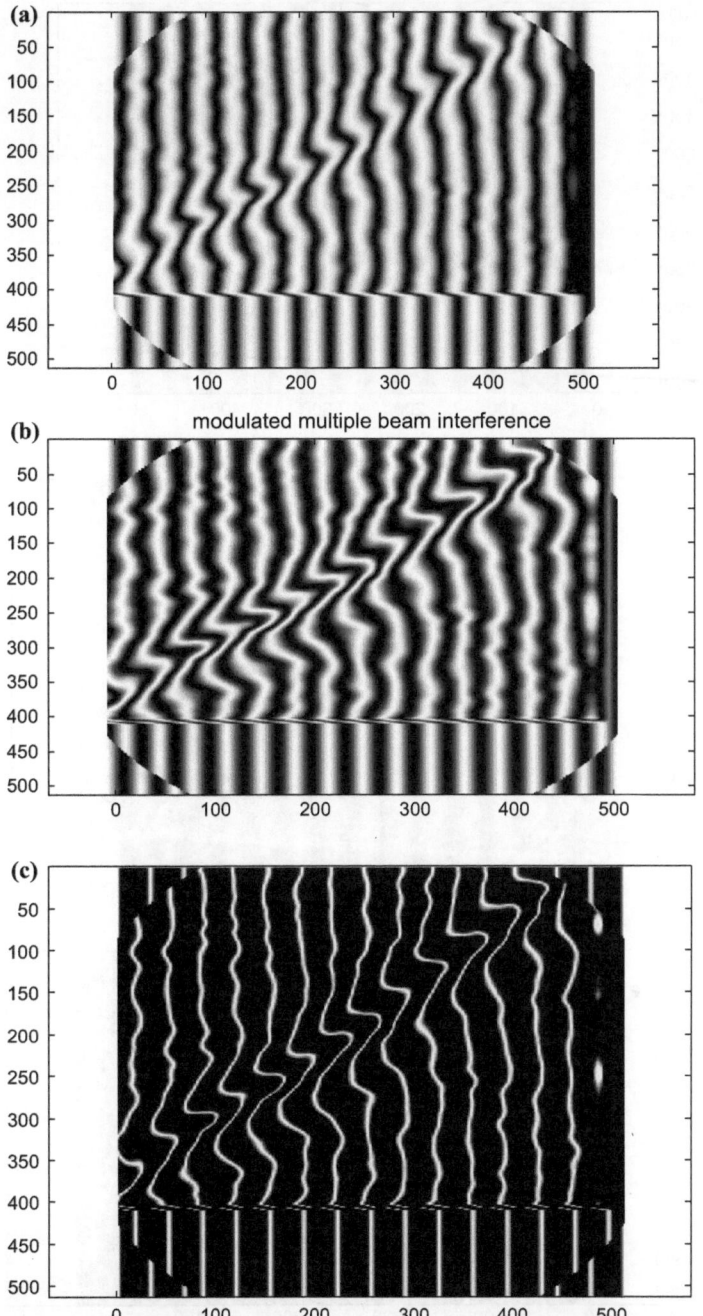

Fig. 6.7 a Modulated ordinary two-beam interference of retinal artery 3. **b** Modulated multiple beam interference of retinal artery 3. **c** Two-beam interference of higher power $\cos^{2N}$, where $N =$ 25. The discontinuous red line is placed on the center of the fringe shift of the segment from artery 3

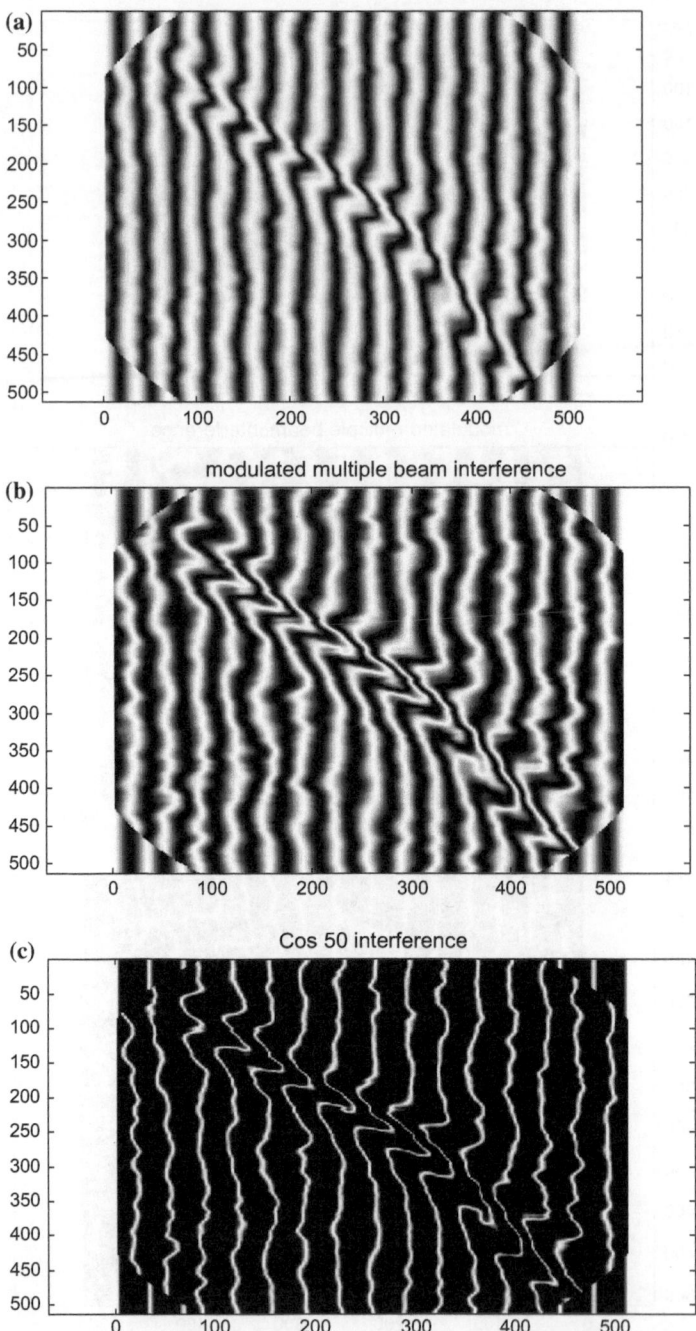

Fig. 6.8   a Modulated ordinary two-beam interference of retinal artery 4. **b** Modulated multiple beam interference of retinal artery 4. **c** Two-beam interference of higher power $\cos^{2N}$, where $N =$ 25. The discontinuous red line is placed in the center of the fringe shift of the segment from artery 4

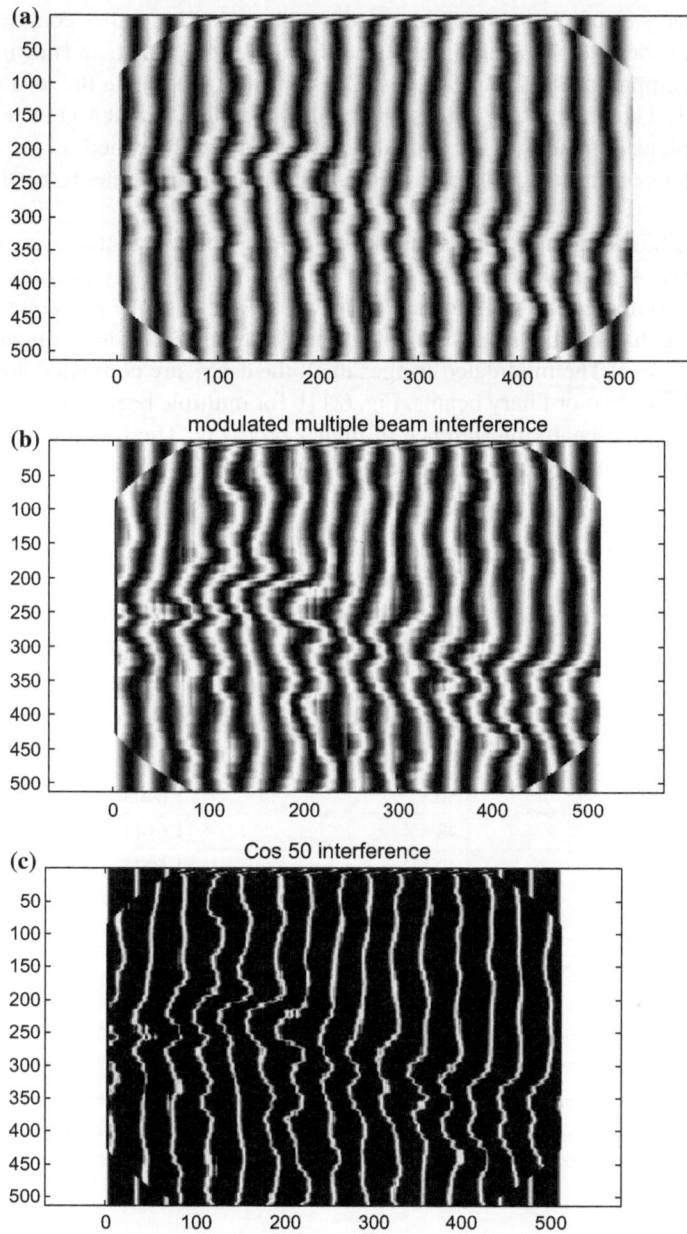

Fig. 6.9 **a** Modulated ordinary two-beam interference of retinal artery 5. **b** Modulated multiple beam interference of retinal artery 5. **c** Two-beam interference of higher power $\cos^{2N}$, where $N =$ 25. The fringe shift of artery 5 is shown

The refractive index corresponding to the segment of artery 1 is computed from Eq. (6.5) considering the mean value of a $(x, y) = 118/256 = 0.461$, and the interfringe spacing is computed as $\Delta Z = 36$ pixels. Consequently, we obtain the results shown in Table 6.1. The error in the values obtained in the table is much greater than the expected values obtained for the retinal refractive index. This method provides the deviation path of the fringe shift, providing information about the refractive index trajectory.

A plaque blocking the retinal arteriole outlined by the horizontal arrow is shown in Fig. 6.10a. A segment from the image with the retinal artery plaque shown as transparent compared with the lower normal artery indicated in the same image is plotted in Fig. 6.10b. The images shown in a and b are resized to have dimensions of 512×512 pixels. The modulated fringes in all the cases, are computed and plotted in Fig. 6.11a for two ordinary beams, Fig. 6.11b for multiple beam interference, and Fig. 6.11c for two-beam interference of higher orders N. The interference patterns corresponding to the segmented image shown in Fig. 6.10b are plotted in Fig. 6.12a.

Table 6.1 Refractive index variation for the retinal segment of artery1 where $\Delta Z = 36$ pixels and the mean value $< a (x, y) > = 118/2$ 56 = 0.461

Z (pixels)	Z image (pixels)	$\delta Z = Z\text{-}Z$ image (pixels)	$\mu(Z) = 1 + < a > \delta Z/\Delta Z$
54	10	44	1.5634
87	46	31	1.3970
121	73	48	1.6147
155	105	50	1.6403
190	140	50	1.6403
224	176	48	1.6147
257	208	49	1.6275
292	241	51	1.6531
326	274	52	1.6659
359	311	48	1.6147
395	347	48	1.6147
428	382	46	1.5891
462	417	45	1.5763
497	449	48	1.6147

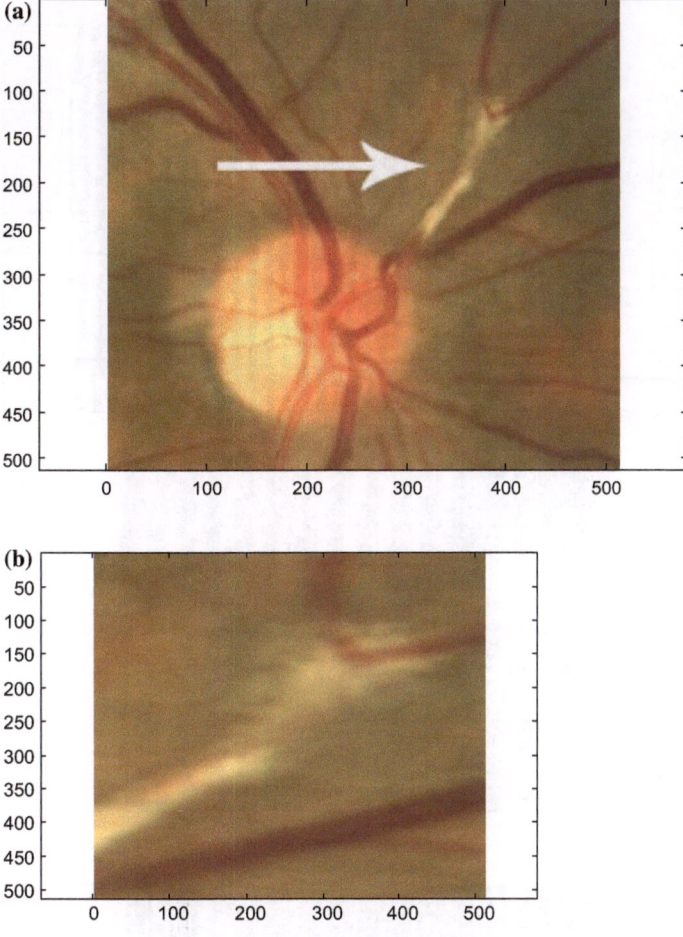

Fig. 6.10 **a** Whole image showing a plaque blocking the retinal arteriole outlined by the horizontal arrow. **b** A segment from the image with the retinal artery plaque shown as transparent compared with the lower normal artery indicated in the same image

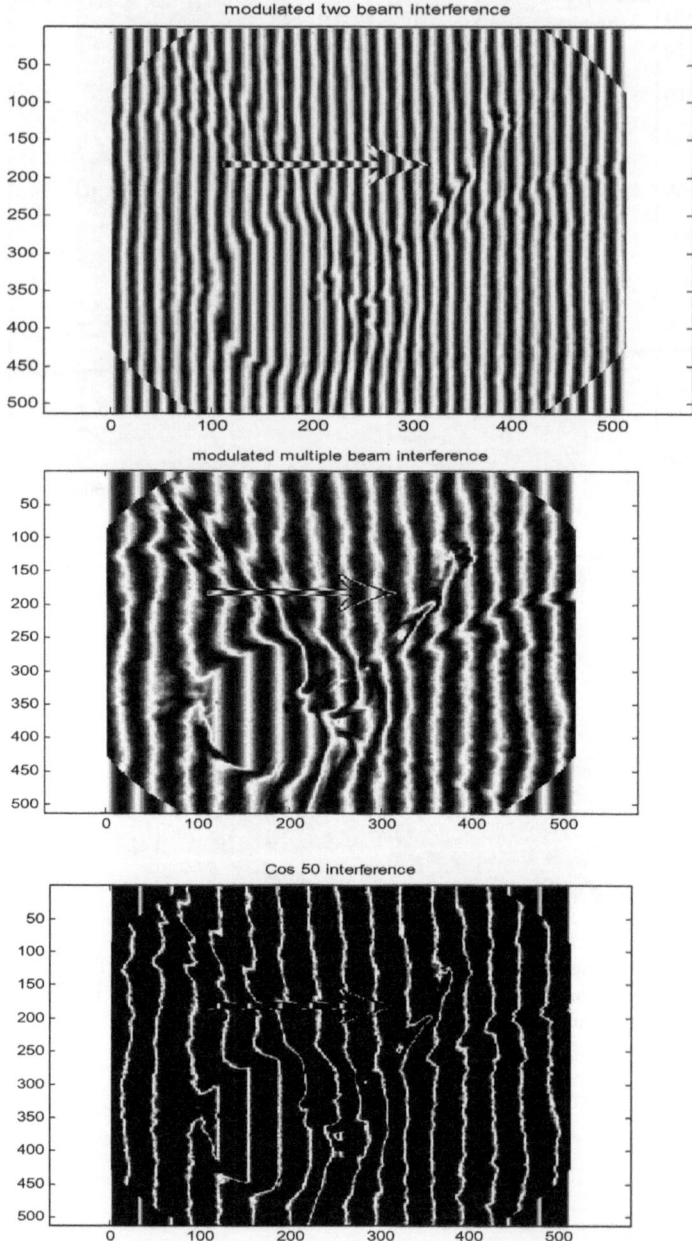

Fig. 6.11 Different interference patterns showing the plaque shift shown in Fig. 6.10a

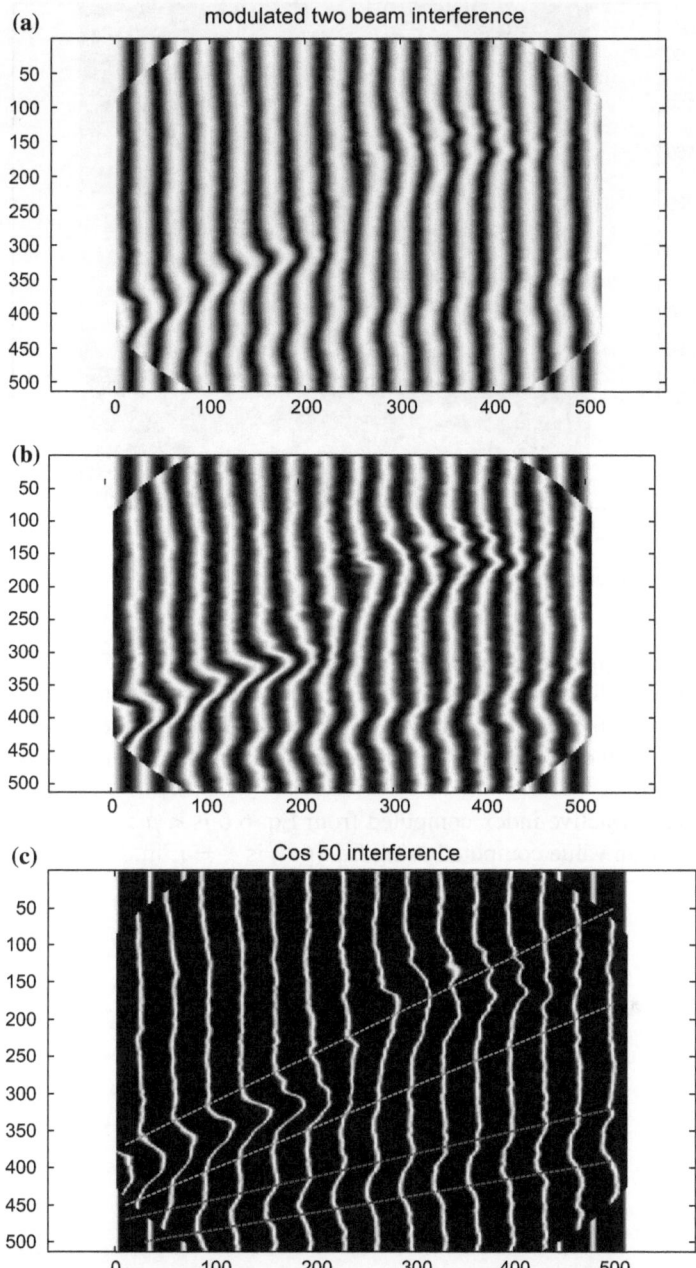

Fig. 6.12 **a** Modulated ordinary two-beam interference of retinal artery plaque shown in Fig. 6.10b.
b Modulated multiple beam interference of retinal artery plaque shown in Fig. 6.10b. **c** Two-beam
interference of higher power $\cos^{2N}$, where $N = 25$. The discontinuous red lines are placed on the
fringe shift of the normal segment from the artery. The discontinuous green lines contour the plaque
segment from the artery

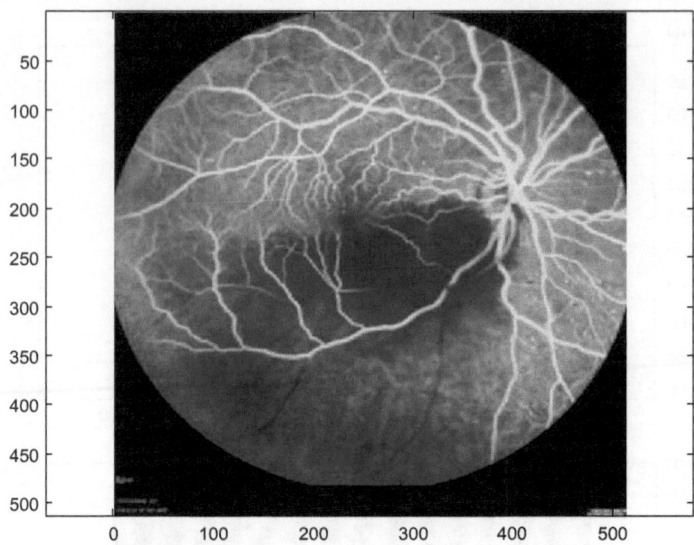

Fig. 6.13 Image of retinal artery occlusion

An image of retinal artery occlusion with dimensions of 512×512 pixels is plotted in Fig. 6.13. The different modulated interference images are computed and plotted as shown in Fig. 6.14a for ordinary two-beam interference, and in Fig. 6.14b for multiple beam interference, and the improved sharper interference image is shown in Fig. 6.14c.

The mean refractive index computed from Eq. 6.6 is $< \mu > = 1.6019$, and the root square mean value computed using Eq. (6.6) is $\sigma = r.$ m. s. $= 2.6447 \times 10^{-2}$.

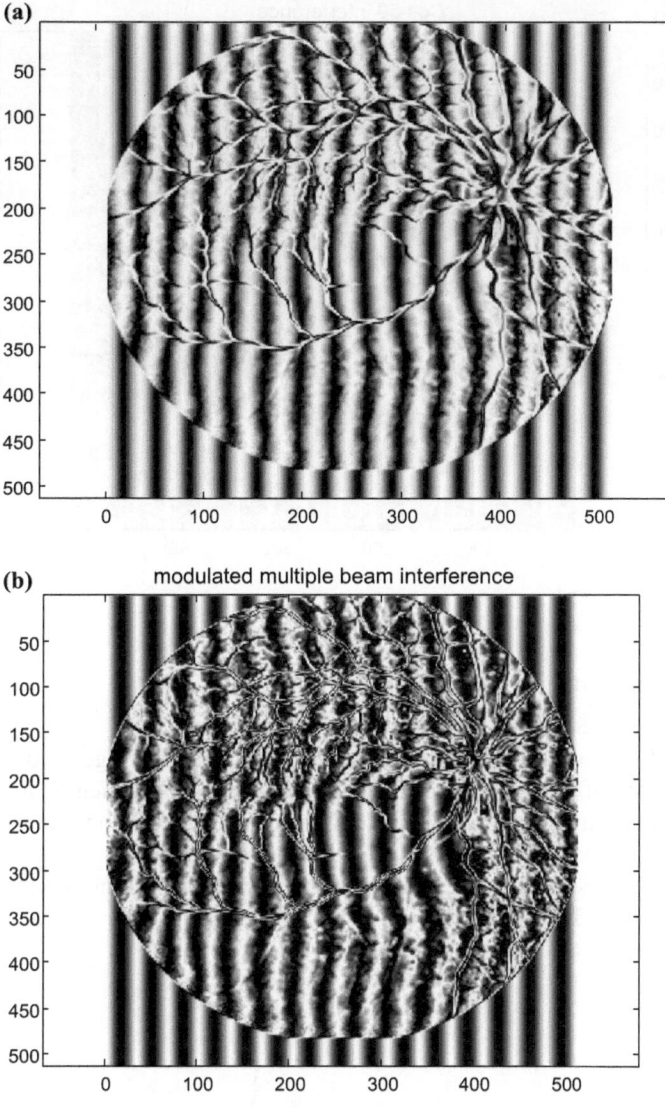

Fig. 6.14 a Modulated ordinary two-beam interference of retinal artery occlusion shown in Fig. 6.13. **b** Modulated multiple beam interference of retinal artery occlusion shown in Fig. 6.13. **c** Two-beam interference of higher power $\cos^{2N}$, where $N = 25$ for the retinal artery occlusion image shown in Fig. 6.13

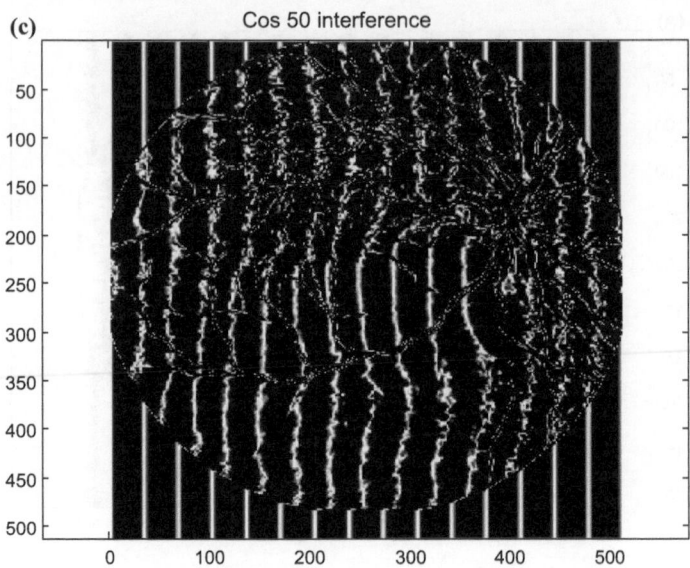

Fig. 6.14 (continued)

6.4 Conclusion

The fringe shift that occurs in the interference straight line fringes is dependent on the optical path difference that occurs from the object. Consequently, the refractive index originating from the optical path difference is deduced from the fringe shift.

The application is the diagnosis of retinal artery occlusion using the interferometer mapping technique. This can be demonstrated by examining the refractive index distribution. The nonuniformity of the fringe shift may be considered an early indicator of retinal arterial occlusion.

References

1. https://webvision.med.utah.edu/book/part-i-foundations/simple-anatomyof-the-retina/
2. N. Pattona, T.M. Aslam, T. MacGillivray et al., Retinal image analysis: CONCEPTS, applications, and potential. Prog. Retin. Eye Res. **25**, 99–127 (2006)
3. A. Frame, M. McCree et al. Structural analysis of retinal vessels, in *Proceedings of the Sixth International Conference on Image Processing, and its applications*, vol. 2 (IEEE, Dublin, 1996), pp. 824–827
4. G.T. Dorner, E. Polska et al., Calculation of the diameter of the central retinal artery from noninvasive measurements in humans. Curr. Eye Res. **25**, 341–345 (2002)
5. M.J. Dumskyj, N. Ishii et al., The accurate assessment of changes in retinal vessel diameter using multiple frame electrocardiograph synchronized fundus photography. Curr. Eye Res. **15**, 625–632 (1996)

6. A.M. Eaton, D.L. Hatchell et al., Measurement of retinal blood vessel width using computerized image analysis. Invest. Ophthalmic. Vis. Sci. **29**, 1258–1264 (1988)

7. X.W. Gao, A. Bharath et al., Quantification and characterization of arteries in retinal images. Comput. Meth. Progr. Biomed. **63**, 133–146 (2000)

8. G. George, M. Wolbarsht et al., Reproducible estimation of retinal vessel width by computerized micrdensitometry. Int. Ophthalmol. **14**, 89–95 (1990)

9. H. Robert Kelly, MD, 929 College Avenue Fort Worth, TX 76104

10. G.E. Sanborn, L.E. Magargal, Arterial obstructive disease of the eye, in *Duane's Clinical Ophthalmology*, ed. W. Tasman, E.A. Jaeger, P.A. (Lippincott Williams & Wilkins, Philadelphia, PA) (2013)

11. J.S. Duker, Retinal arterial obstruction, in *Ophthalmology*. 4th ed, ed. M. Yanoff, J.S. Duker (Elsevier, Philadelphia, PA), chap. 6.18 (2014)

12. R.K.Z. Ma, F. Wang et al., High-speed spectral domain optical coherence tomography for imaging of biological tissues, in *Proceedings of the Optics in Health Care and Biomedical Optics: Diagnostics and Treatment II of Proceedings of SPIE*, ed. by B. Chance, M. Chen, A.E.T. Chiou, Q. Luo (Beijing, China, 2004), pp. 286–294

13. M. Wojtkowski, R. Leitgeb et al., In vivo human retinal imaging by Fourier domain optical coherence tomography. J. Biomed. Opt. **7**(3), 457–463 (2002)

14. J.F. De Boer, B. Cense et al., Improved signal-to-noise ratio in spectral-domain compared with time-domain optical coherence tomography. Opt. Lett. **28**(21), 2067–2069 (2003)

15. O. Puzyeyeva, W. Ching Lam et al., High-resolution optical coherence tomography retinal imaging: a case series illustrating potential and limitations. J. Ophthalmol. 1–6 (2011)

16. A.M. Hamed, Image processing of Corona virus using interferometry. Opt. Photon. J. **6**, 75–86 (2016)

Chapter 7
Investigation of Kidney Images Using Cascaded Fabry–Perot Interferometer (CFPI)

In this paper, a new multiple beam interference setup using a cascaded Fabry–Perot Interferometer (CFPI) is considered. The application of this CFPI to kidney images is described. The sharp fringes are deformed by the kidney phase, giving modulated fringes. The sharp fringe shift of the kidney images obtained in the case of the multipass CFPI model is investigated. In addition, the image contrast is compared with the contrast in the case of ordinary FPIs. Finally, the modulated two-beam interference of kidney images is given for comparison.

7.1 Introduction

The idea of multipassed FPI was initially suggested in [1]. Later, a double Fabry–Perot interferometer was designed to improve the fringe contrast providing nearly 30 times the contrast of a single FPI [2].

Image processing of biomedical images is useful in the investigation of images, particularly for ultrasonic medical images provided with multiplicative speckle noise filtering, which is important for eliminating noise using, for example, Wiener or median filtering. The mean filter is not useful for multiplicative noise.

The development of a semiautomated program that uses image processing techniques and geometric principles to define the boundary, and segmentation of the kidney area, and to enhance kidney stone detection [3–5].

The discrete wavelet transform (DWT) is attractive for denoising since it has the advantage of varying the scale at which the component frequencies are analyzed [6–8] compared with Fourier transform techniques. Hence, speckle noise, which is multiplicative decreases, increasing the S/N ratio in the DWT compared with the corresponding S/N ratio in the case of the Wiener and median filters. Recently, as outlined in ref. [9], wavelet denoising involves the following stages: calculating the DWT, removing noise by changing the wavelet coefficients, and applying the

© The Author(s), under exclusive license to Springer Nature Switzerland AG 2024   85
A. Hamed, *Cascaded Interferometers and Their Medical Applications*,
SpringerBriefs in Applied Sciences and Technology,
https://doi.org/10.1007/978-3-031-64535-8_7

inverse (IDWT) to construct the despeckled image. Another method of improving the ultrasonic low contrast of kidney images is based on the reduction of speckle noise using the Gabor filter; then, the despeckled image is enhanced using histogram equalization [10]. Image processing of some biomedical images is performed using two-beam and ordinary multiple beam techniques [11–13], leading to extraction of phase information from the images.

In this chapter, sharp interferometric images produced from the sequence of interferometers using the CFPI compared with the ordinary FPI are shown. The phase information of kidney images obtained from the modulated interference images is obtained.

In a recent publication by the author [14], we considered multiple passes of two-beam interference, while in this study, we considered CFPI arranged in series.

7.2 A Cascaded Fabry–Perot Interferometer (CFPI)

A higher-order multiple beam interference composed of four cascaded interferometers is schematically represented in Fig. 7.1. A He-Ne laser beam is spatially filtered and rendered parallel using an objective lens L followed by a pinhole P placed in the short focus of the objective lens, and a converging lens L_1 placed at the focal plane f from the pinhole. The collimated laser beam passes through the four F.P.I.s arranged in series followed by the Fourier transform lens L_2 of focal length f_2. The Fourier and imaging planes are located as shown in Fig. 7.1.

The transmitted intensity distribution in the case of an ordinary FPI is given by the following formula [11]:

$$I(\delta; R) = \frac{T^2}{1 + R^2 - 2R\cos(\delta)} \tag{7.1}$$

where T is the transmission coefficient and R is the reflection coefficient of the interferometer. δ, is the phase difference between any two adjacent emerging rays.

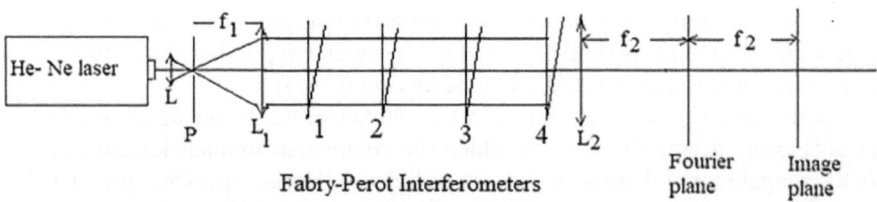

Fig. 7.1 Multiple beam interferometer composed of four cascaded interferometers. L objective lens, P pinhole, and L_1 converging lens where the elements L, P, and L_1 render the laser beam spatially filtered and collimated. L_2 is the Fourier transform lens of focal length f_2. The Fourier and imaging planes are shown in the figure

In the case of cascaded interferometers, the intensity distribution is an ordinary distribution to a power of N, where N is the number of cascaded interferometers. Then, the intensity distribution is represented as follows:

$$I(\delta; R, N) = \left[\frac{T^2}{1 + R^2 - 2R\cos(\delta)} \right]^N \tag{7.2}$$

The maximum intensity is computed as:

$$I(\delta = 2\pi; R, N) = I_{max}(R, N) = \left[\frac{T^2}{(1-R)^2} \right]^N \tag{7.3}$$

Using Eqs. (7.2) and (7.3), the normalized intensity due to cascaded multiple beam interference can be written as follows:

$$I_{normalized}(\delta; R, N) = \frac{I}{I_{max}} = \frac{1}{\left[1 + F\sin^2\left(\frac{\delta}{2}\right)\right]^N} \tag{7.4}$$

Where $F = \dfrac{4R}{(1-R)^2}$ is the coefficient of finesse. $\tag{7.5}$

F is a measure of fringe sharpness and contrast.

For greater values of reflectivity, $R > 80\%$, F is much larger than one; hence, an approximate expression for the intensity is obtained as:

$$I_{normalized}(\delta; R, N) \sim 1 \left/ \left[F\sin^2\left(\frac{\delta}{2}\right) \right]^N \right. \tag{7.6}$$

7.3 Results and Discussion

The normalized coefficient of finesse F as a function of reflectivity R for different cascaded interferometers is computed from equation (7.6) and represented as shown in Chap. 5, Fig. 5.2. $N = 1$ represents the ordinary FPI, while $N = 2, 3$, and 4 represent the number of cascaded interferometers.

On the left, is a normal kidney image, while on the right, a polycystic kidney image is shown in Fig. 7.2. This image is extracted from the following site: www.kidneyfailure.com. Kidney failure was investigated using two-beam interference. On the left side, a nearly regular shift is shown for the normal kidney, while an irregular random shift is obtained on the right side corresponding to kidney failure, as shown in Fig. 7.3a, where $M = 15$. The details of the irregular shift for kidney failure are shown using a greater number of interference fringes, $M = 60$, as shown in Fig. 7.3b.

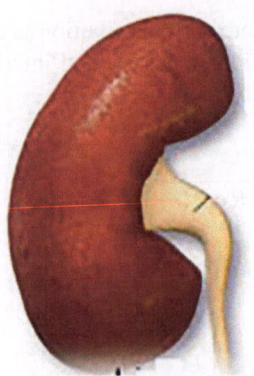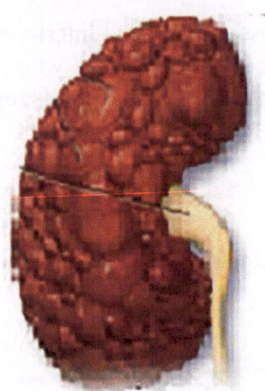

Fig. 7.2 On the left, is a normal kidney image, while on the right, a polycystic kidney is shown. This image is extracted from the following site: www.kidneyfailure.com

A segment from the normal kidney modulated by two-beam interference is shown in Fig. 7.4a, while a segment of failed kidney modulated by two-beam interference is shown in Fig. 7.4b.

Another plot of both kidneys in which failure occurred, as shown in Fig. 7.5, revealed random interference within the kidney, as shown in Fig. 7.6. A plot of normal and cancerous kidneys is shown in the Fig. 7.7a, b. A normal kidney modulated by two-beam interference fringes is shown in Fig. 7.8a, while the cancerous kidney modulated by interference showed a contour encircling the cancerous part, as shown in Fig. 7.8b. The modulated cancerous part of the kidney is shown in Fig. 7.8c.

The ordinary F P I where the number of interferometers $N = 1$, is shown in the Fig. 7.9a. The cancerous kidney shown in the Fig. 7.7b is modulated by 30 fringes. The cascaded higher orders of F P I that modulated the cancerous kidney are shown in the Fig. 7.9b, c, d. In all the figures, the irregular segment shown in the upper part of the kidney represents the cancerous part. The cascaded arrangement of F P I modulated by the image is computed from Eq. (7.5), where the image matrix is introduced in the modulation interference phase term.

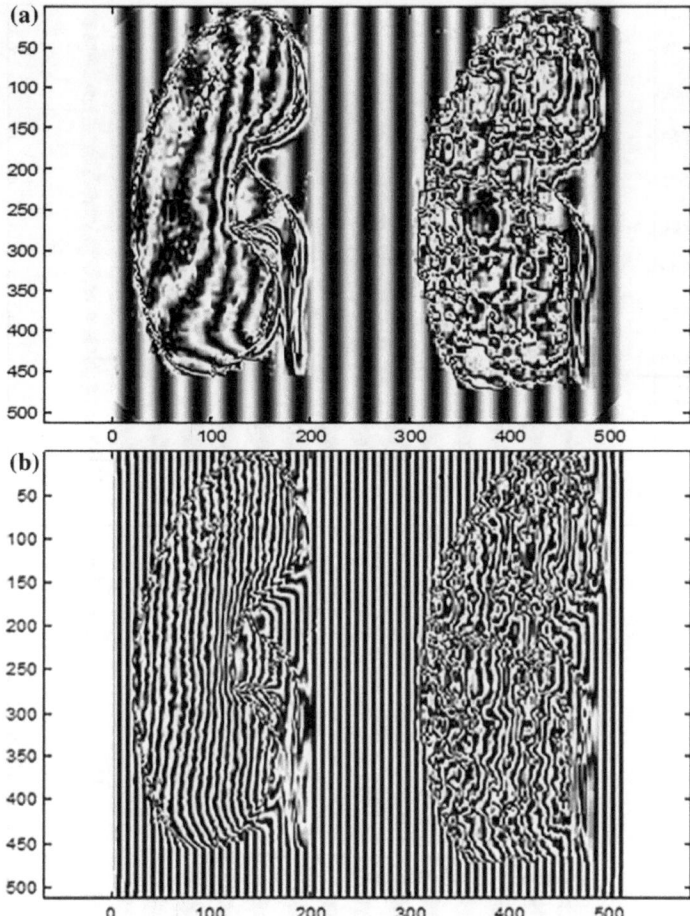

Fig. 7.3 **a** Kidney failure is investigated using two-beam interference. On the left side, a nearly regular shift is shown for the normal kidney, while an irregular random shift is obtained on the right side corresponding to kidney failure. The number of fringes $(M) = 15$. **b** Kidney failure is investigated using two-beam interference. On the left side, a nearly regular shift is shown for the normal kidney, while an irregular random shift is shown on the right side corresponding to kidney failure. Number of fringes $(M) = 60$

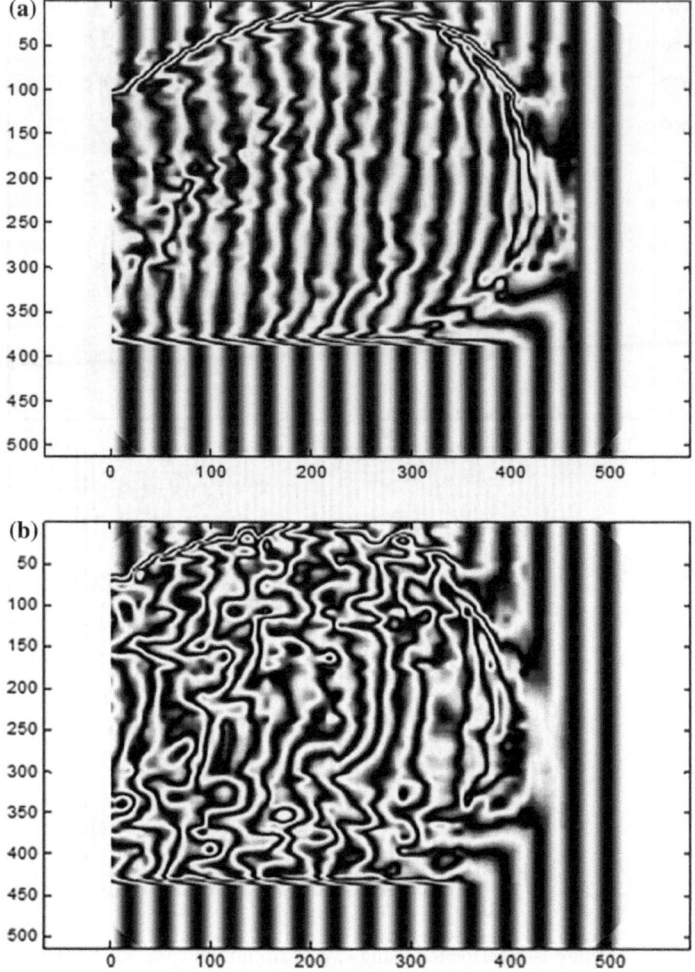

Fig. 7.4 **a** A segment of a normal kidney modulated by two-beam interference. **b** Segment of the failed kidney modulated by two-beam interference

Fig. 7.5 The figure shows
that both kidneys failed

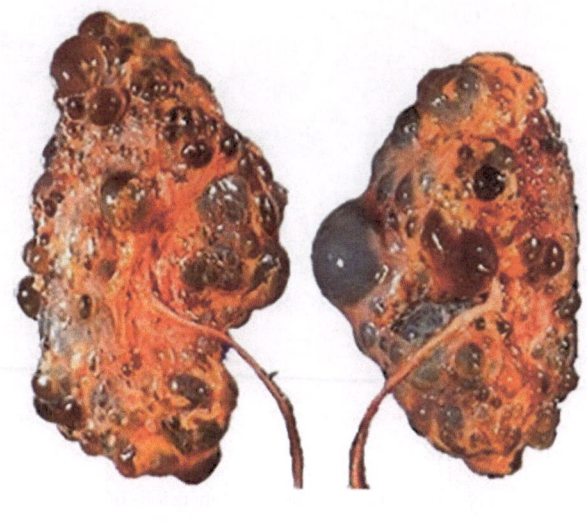

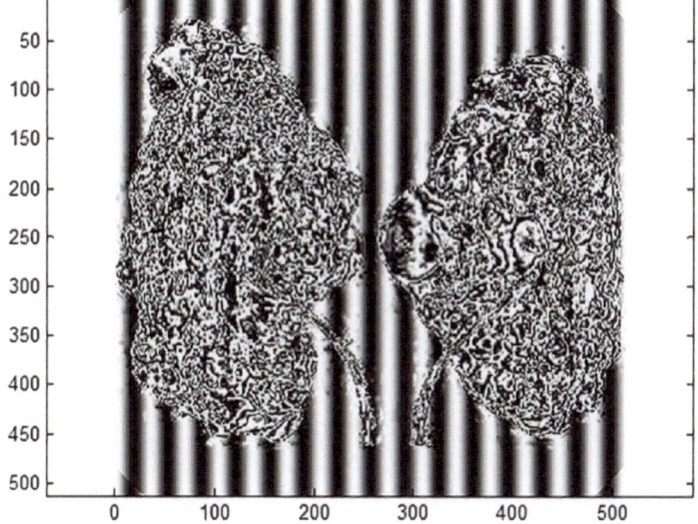

Fig. 7.6 It is shown in the modulated interference that both kidneys have a failure outlined by randomness in the interference pattern

(a)   (b)

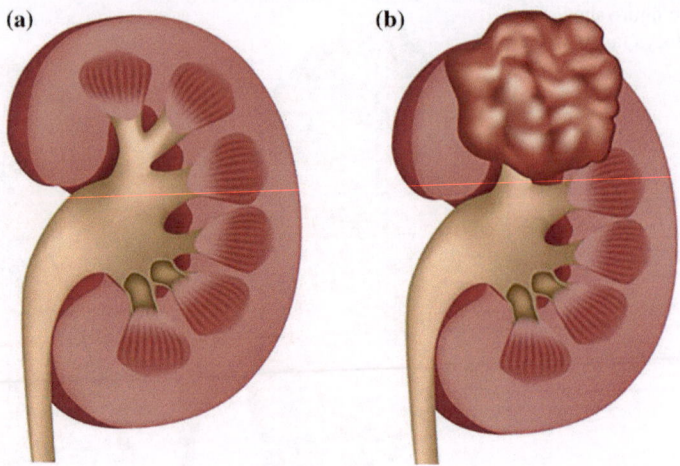

Fig. 7.7 **a** Normal kidney image with dimensions of 552 × 438 pixels. **b** Cancerous kidney image with dimensions of 552 × 438 pixels

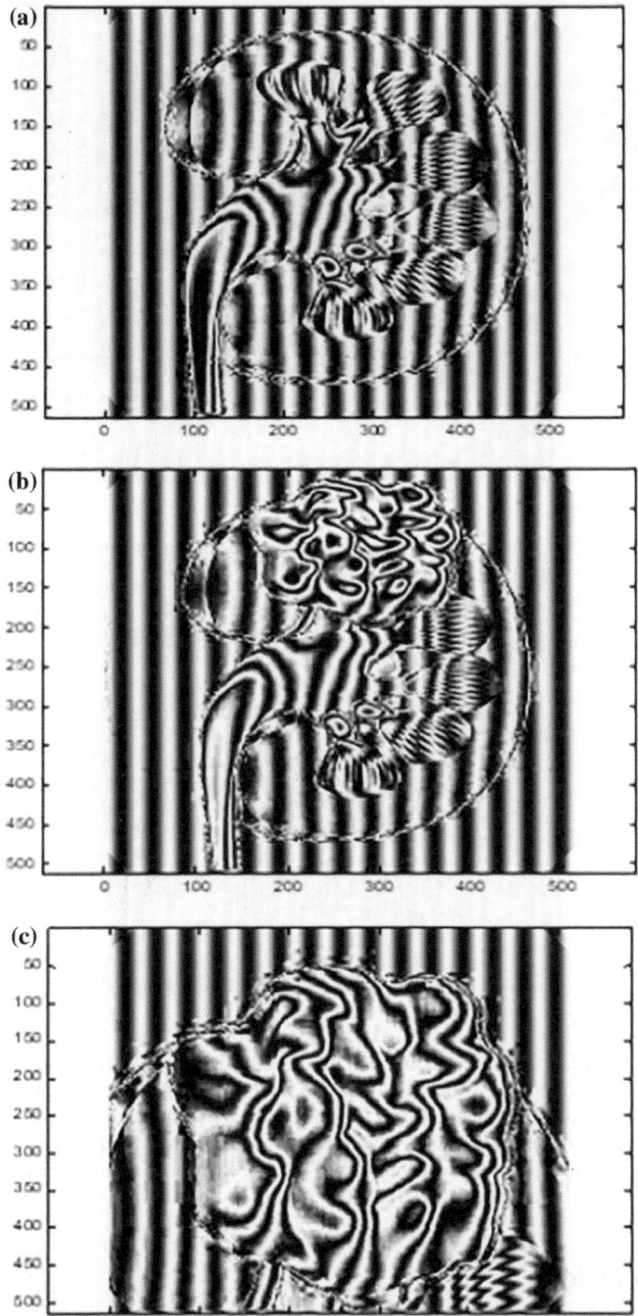

Fig. 7.8 **a** Normal kidney modulated by two-beam interference fringes. **b** Contour fringes are shown in the upper part of the cancerous kidney. **c** The modulated cancerous part of the kidney

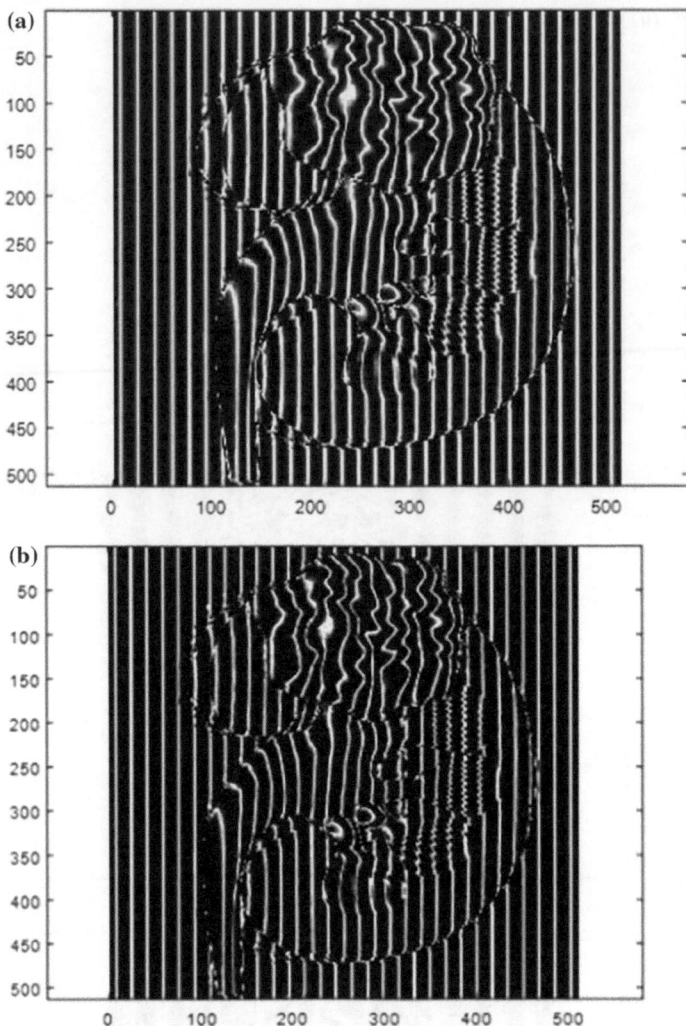

Fig. 7.9 **a** Ordinary F P I where the number of interferometers $N = 1$. The cancerous kidney shown in Fig. 7.7b is modulated by 30 fringes. **b** Cascaded F P I where the number of interferometers is $N = 2$. The cancerous kidney shown in Fig. 7.7b is modulated by 30 fringes. **c** Cascaded F P I where the number of interferometers is $N = 3$. The cancerous kidney shown in Fig. 7.7b is modulated by 30 fringes. **d** Cascaded F P I where the number of interferometers is $N = 4$. The cancerous kidney shown in Fig. 7.7b is modulated by 30 fringes

7.4   Conclusion

We suggest an improved CFPI using the cascaded arrangement of multiple beam interferometers. The contrast is further improved compared with that of the ordinary FPIs. In addition, a comparison of the interferometric images with the two-beam

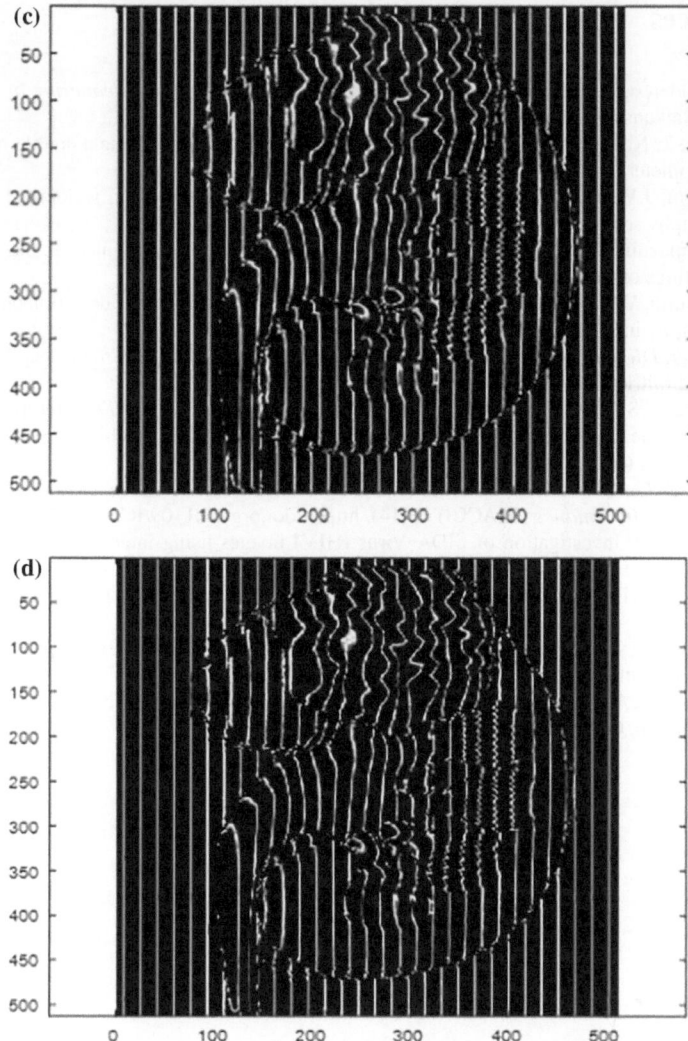

Fig. 7.9 (continued)

interference is shown. Kidney images that were either normal or cancerous showed regular interference fringes in the case of normal kidneys, while cancerous kidney contours and irregular shapes are shown in two and multiple beams, respectively. The coefficient of finesse as a function of reflectivity is plotted for ordinary and cascaded F P I.

References

1. J.R. Sandercock, in *Procceding 2nd International Conference Light Scattering in Solids*, ed. by M. Balkanski (Flammarion sciences, Paris, 1971), pp. 9
2. S. Itoch, T. Nakamura, Brillouin scattering study of $Gd_2(MoO_4)_3$ using a double fabry-perot interferometer. Japanese J. Appl. Phys. **14**, 83–86 (1975)
3. D.Y. Kim, J.W. Park, Computer-Aided detection of kidney tumor on abdominal computed tomography scans. Acta Radiologic **45**(7), 791–795 (2004)
4. J. Kolomaznik, Fast segmentation of kidneys in CT images, presented at the WDS '10 Proceedings of Contributed Papers, pp. 70–75 (2010)
5. S. Ebrahimi, V.Y. Mariano, Image quality improvement in kidney stone detection on computed tomography images. J. Image Graph. **3**(1) (2015)
6. W. Fisher, *Digital Television* (Springer-Verlag, A Practical Guide for Engineers, 2004)
7. S. Papadimitriou, A. Bezerianos, J. Syst. Architect. **42**, 55–65 (1996)
8. A. Khare, U.S. Tiwary, Int. J. Wavelets Multiresolut. Inf. Process. **3**, 477–496 (2005)
9. C.M. Nicolae, L. Moraru, Annals of the University of Craiova. Math. Comput. Sci. Series **38**(1), 27–34 (2011)
10. V. Kala, R. Gunasundari, In: *International Conference on Advances in Computing, Communications, and Informatics* (ICACCI) (2014). https://doi.org/10.1109/ICACCI.2014.6968485
11. A.M. Hamed, Investigation of SIDA Virus (HIV) images using interferometry and speckle techniques. Int. J. Innovative Res. Comput. Sci. Tech. (IJIRCST) **4**, 38–45 (2016)
12. A.M. Hamed, Image processing of coronavirus using interferometry. Opt. Photonics J. **6**, 75–86 (2016)
13. A.M. Hamed, Topics on optical and digital image processing using holography and speckle techniques, published by Lulu.com, ISBN 9781329328464 (2015)
14. A.M. Hamed, A modified michelson interferometer and an application on microscopic imaging. Int. J. Photon. Optic. Technol. (IJPOT) **3** (2017)

Chapter 8
Coronavirus Image Processing Using Interferometry

A new method for image processing of coronaviruses based on two and multiple beam interference is suggested. The method is based on measuring the fringe shift to the background interference pattern. The interesting application of the coronavirus image in confocal microscopy is obtaining depth information since it has the property of optical sectioning. An accurate measurement of the fringe shift is obtained using multiple beam interference since the contrast is greater than that for two-beam interference. The refractive index of the corona virus image is deduced from the fringe shift. A MATLAB code is used for the processing of all images.

8.1 Introduction

Coronaviruses are an important family of human and veterinary pathogens that can cause enteric and respiratory infections. Coronavirus infection can lead to respiratory failure, gastroenteritis, nephritis, and hepatitis.

A novel methodology of single particle image analysis is applied to select virus features to obtain a detailed model of the oligomer state and spatial relationships among viral structural proteins [1].

The addition of electronics, computers, and software to interferometry has enabled enormous improvements in optical metrology. Phase-shifting interferometry is used for getting data into a computer so that the data can be analyzed [2–9]. Image processing of uniform objects and modified apertures was outlined [10–14].

In this paper, the refractive index distribution of coronavirus images is computed via the phase shift method. The coronavirus fringe shift to background interference is computed to obtain useful information about the phase shift of the image leading map of the height depth and the refractive index distribution. The results and discussion are given, followed by a conclusion. The former work concerning the digital fringe shift is

© The Author(s), under exclusive license to Springer Nature Switzerland AG 2024   97
A. Hamed, *Cascaded Interferometers and Their Medical Applications*,
SpringerBriefs in Applied Sciences and Technology,
https://doi.org/10.1007/978-3-031-64535-8_8

limited by $\lambda/2$, the interfringe spacing. Hence, the computation of the refractive index is dependent on the fringe spacing using either two or multiple beam interference.

8.2   Analysis

The complex amplitude of the coronavirus can be represented as follows:

$$A_{object}(x, y; z) = a \exp[j\phi(x, y; z)] \tag{8.1}$$

where a is the amplitude of the image, and $\Phi(x, y; z)$ is its phase for object depth z. Equation (8.1) can be written in a discrete matrix form as follows:

$$A_{object}(x, y; z) = \sum_{n=1}^{N} \sum_{m=1}^{M} a \exp\left[j\phi(n\Delta x, m\Delta y; z)\right] \tag{8.2}$$

where a square matrix of dimensions $N \times N = 512 \times 512$ pixels $(N = M)$ is assumed.

This chapter focuses on the main technique for phase evaluation of coronavirus images using the phase-shifting method.

The coherent addition of a reference laser beam $A_r = R \exp[j\psi(x, y)]$ to the above object beam is considered to fabricate the modeled interference pattern. The laser beam is spatially filtered using a pinhole located in the focal plane of a converging lens to obtain uniform illumination. In this case, a plane wave is obtained. The pinhole passes only the central peak from the whole diffraction pattern and suppresses all the diffraction legs, and then the converging lens, which is placed at a distance (f) from the pinhole, passes parallel rays of uniform intensity, which is considered a plane wave.

Then, the intensity of the two-beam interference obtained in the detector plane can be expressed as the modulus squared as follows:

$$I(x, y, z) = I_0\{1 + M \cos[\phi(x, y; z) - \psi(x, y)]\} \tag{8.3}$$

where $I(x, y, z)$ is the intensity of the modulated interference field at point (x, y) for an object depth z, $I_0 = R^2 + a^2$ is the function that characterizes the mean intensity of the interference pattern and $M = \frac{2aR}{R^2+a^2}$, is the function that determines the modulation of the interference signal. In this case, the obtained trigonometric function has straight-line fringes modulated by the object phase information. The modulated intensity is rewritten in matrix form as follows:

$$I(x, y, z) = I_0 \sum_{n=1}^{N} \sum_{m=1}^{M}\{1 + M \cos[\phi(n\Delta x, m\Delta y; z) - \psi(n\Delta x, m\Delta y)]\} \tag{8.4}$$

Certainly, for the d.c. term in Eq. (8.3), $I_0 = a^2 + R^2$.

R is the amplitude of the coherent laser beam, and the phase of ψ appears in Eq. (8.3). Equation (8.4) is used in the fabrication of the phase-shifted images outlined in Eq. (8.5).

Since the distance between any two fringes $= \lambda/2$. Consequently, according to the phase shift technique [7, 9], the phase information of the image is governed by the following equation:

$$\psi(n\Delta x, m\Delta y) = \phi(n\Delta x, m\Delta y; z) - \tan^{-1}\left[\frac{I_3(x, y) - I_2(x, y)}{I_1(x, y) - I_2(x, y)}\right] \quad (8.5)$$

where the range of the interference phase $(n\Delta x, m\Delta y)$ extends from 0 to 2π for a height z, I_1 is the intensity given in Eq. (8.4) at a phase $\Psi = \pi/2$, I_2 has $\Psi = \pi$, and I_3 has phase $\Psi = 3\pi/2$. Then, three equations are solved to obtain Eq. (8.5).

Once the phase is determined across the interference field, the corresponding height distribution $h(x, y)$ on the surface of the coronavirus can be determined [2] as follows:

$$h(x, y; z) = \frac{\lambda}{4\pi}\phi(n\Delta x, m\Delta y; z) \quad (8.6)$$

We assumed that the surface was measured at normal incidence. Almost all interferometers used to measure surface height variations use phase-shifting techniques.

The refractive index of the corona virus μ is computed as follows:

Since the phase of the wave cumulates traveling a distance L in a medium is

$$\phi(x, y) = \int_L k\,dl = \int \frac{\mu(x, y, z)\omega}{c}\,dl \quad (8.7)$$

Then, the same wave that propagates over two equivalent paths L in the coronavirus medium and vacuum gives the phase difference as shown in Fig. 8.1:

$$\Delta\phi(x, y, z) = \int_L (k - k_0)\,dl = \int [\mu(x, y, z) - 1]\frac{\omega}{c}\,dl \quad (8.8)$$

where $k = \omega/c = 2\pi/\lambda$ is the propagation wavenumber in a medium with refractive index μ and k_0 is the propagation constant in vacuum.

By differentiating to the path $l = z$, the refractive index distribution of the coronavirus image is computed as follows:

$$\mu(x, y) = 1 + \frac{c}{\omega}\frac{d}{dz}\left[\Delta\phi(x, y, z)\right] \quad (8.9)$$

Fig. 8.1 Propagation of light in a medium with refractive index μ compared with that in air

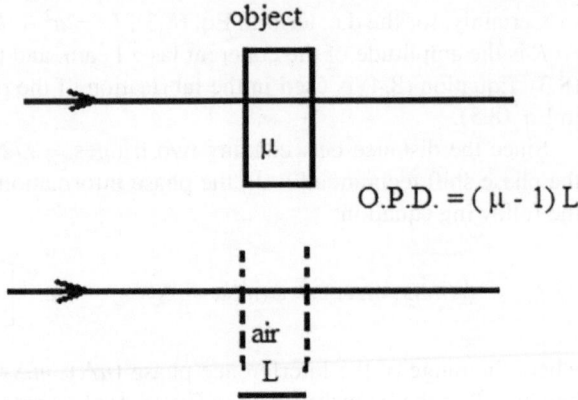

$$O.P.D. = (\mu - 1)\,L$$

Since the angular frequency is related to the wavelength as $\omega = 2\pi c/\lambda$, and $\Delta\phi(x, y) = \frac{2\pi}{\lambda}$ O.P.D.; then, the above equation becomes:

$$\mu(x, y) = 1 + \frac{d}{dz}[O.P.D.] \tag{8.10}$$

The optical path difference represents the height variation of the image, namely, $h(x, y, z)$; then, Eq. (8.10) becomes:

$$\mu(x, y) = 1 + \frac{d}{dz}\big[h(x, y, z)\big] \tag{8.11}$$

The differentiation of the height distribution $h(x, y, z)$ to z gives the differential fringe shift and the amplitude of the planar image $a(x, y)$. Consequently, we finally obtained Eq. (8.12).

$$\mu(x, y)_{const.x} = 1 + a(x, y)\frac{dz}{\Delta z} \tag{8.12}$$

The fringe shift is δZ to the interfringe spacing Δz at constant x, the fringes are assumed to be in the x–y plane, z is the axis normal to the fringe system, which represents the height depth, and $a(x, y)$ represents the amplitude of the image. In Eq. (8.12), $h(x, y, z) = a(x, y).\delta z$.

8.2.1 Computation of Contrast

The visibility expression representing the fringe contrast is as follows:

$$C = \frac{I_{max} - I_{min}}{I_{max} + I_{min}}.$$

Substitute this result in Eq. (8.3) for comparison:

$$C = M = \frac{2R}{1 + R^2} \tag{8.13}$$

The contrast given in the case of multiple beam interference is extracted from the transmitted intensity distribution [15]:

$$I_t = \frac{I_0}{1 + F \sin^2[\phi(x, y) - \psi(x, y)]/2} \tag{8.14}$$

The parameter $F = \frac{4R}{(1-R)^2}$ and the corresponding contrast are given by formula (8.14). The maximum and minimum intensities are obtained from Eq. (8.14) as: $I_{max} = I_0$ for $\phi(x, y) - \psi(x, y) = 2m\pi$, where m is an integer.

$$I_{min} = \frac{I_0}{1 + F} \text{ for } \phi(x, y) - \psi(x, y) = m\pi$$

The refractive index directly related to the polychromatic spectral distribution of illuminating light according to the Cauchy formula as follows:

$$\mu_\lambda = a + b/\lambda^2 \tag{8.15}$$

where a and b are constants. Then, the fringe shift and the refractive index are affected by the change in the wavelength.

8.3 Results and Discussion

The coronavirus image used in the processing is shown in Fig. 8.2. It has dimensions of 512×512 pixels. The fringe shift that occurred within the coronavirus region to the background shift is plotted in Fig. 8.3 at frequency $f = 1/32$. This interferometer plot is obtained from Eq. (8.3) written in discrete form, where $M = N = 512$ pixels. The yellow discontinuous horizontal lines are taken at 10, 80, 140, 200, 260, 320, 380, and 440 pixels. Only 7 fringes are shown in this image. The shift of the coronavirus cells was computed compared with the interfringe spacing. The MATLAB code is used for the computation of the modulation term $\cos(y - (1/32) A (i, j))$. The background two-beam interference is only cos (y). It is represented by straight-line fringes modulated by the object matrix A (M, N). Figure 8.3 shows eight plots with 10, 80, 140, and 200 pixels in the 1st column, and 260, 320, 380, and 440 pixels in the 2nd column. The upper left plot at 10 pixels shows uniform straight-line fringes, which are Fig. 8.4 compared with the shifted fringes corresponding to the image geometry shown at the mentioned lines. Hence, we can obtain phase information about the image, as plotted in Fig. 8.5, using Eq. (8.8) and height information from Eq. (8.9).

Four different modulated fringe shifts are shown in Fig. 8.6. Interferometry images of the coronavirus using the cosine function to represent the phase at four different spatial frequencies: 1/32, 1/64, 1/96, and 1/128. The original image is multiplied by a factor of $\alpha = 1/32$ in the interference-modulated terms as follows:

For (a) cos ($y - (1/32) A(i, j)$), (b) cos ($2y - (1/32) A(i, j)$), (c) cos ($3y - (1/32)$ $A(i, j)$), and (d) cos ($4y - (1/32) A(i, j)$).

Multiple beam interferometry images of the coronavirus using the Airy function to represent the phase at four different background frequencies are plotted in Fig. 8.7. The original image is multiplied by a factor of $\alpha = 1/32$, as shown in Fig. 8.6.

Finally, comparing the fringe shift in the coronavirus image at different frequencies $f = 1/32$, 1/64, 1/96, and 1/128, it is shown that the fringe shift is not resolved at frequencies greater than $f = 1/128$.

The profile of the coronavirus image taken at constant $x = 150$ pixels extracted from the image shown in Fig. 8.3 is plotted in Fig. 8.8a, and the profile taken at $x = 330$ pixels is plotted in Fig. 8.8b. On the left of both plots, seven fringes are shown Fig. 8.8.

The profiles are affected by the fringe shift resulting from the phase change of the image under consideration.

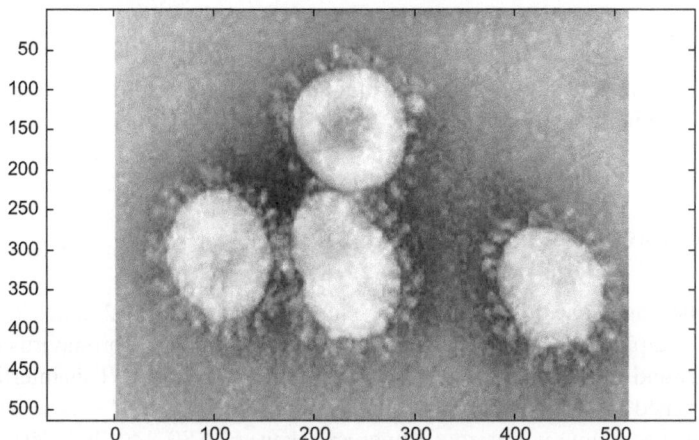

Fig. 8.2 Coronavirus image used in the processing. The matrix has dimensions of 512×512 pixels

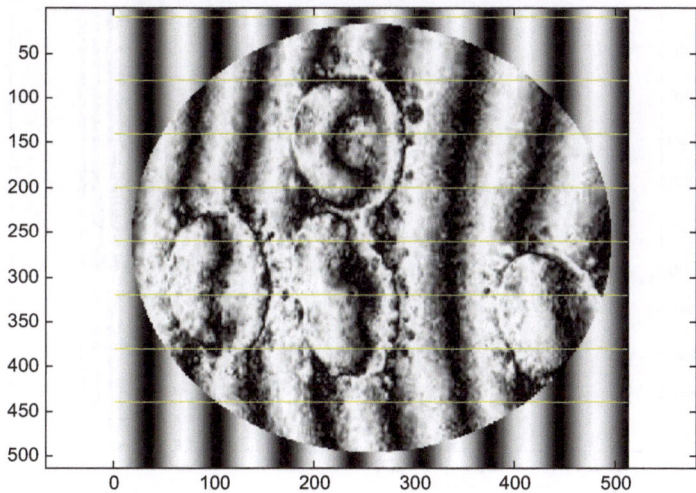

Fig. 8.3 Two beam interference of coronavirus with dimensions of 512×512 pixels. The yellow discontinuous lines are taken at 10, 80, 140, 200, 260, 320, 380, and 440 pixels. Seven fringes are shown in this image. The shift of the corona virus cells is computed compared with the interfringe spacing

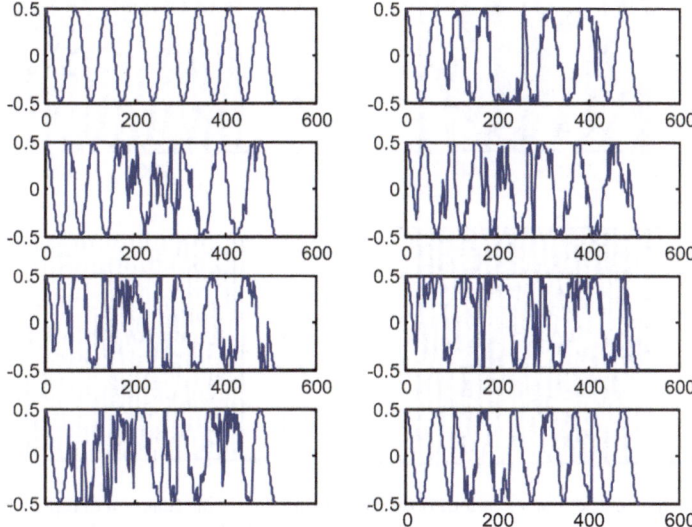

Fig. 8.4 Eight plots from Fig. 8.3 are shown at 10, 80, 140, and 200 pixels in the 1st column on the left, and at 260, 320, 380, and 440 pixels in the 2nd column

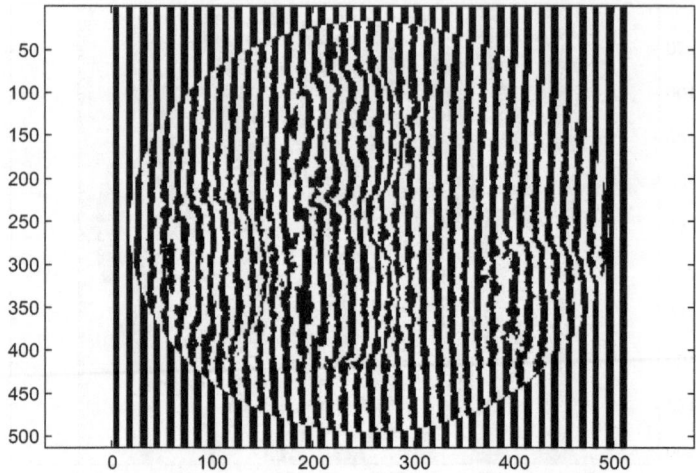

Fig. 8.5 Phase map of the coronavirus with background spatial frequency $f = 5$, where $\alpha = 1/32$

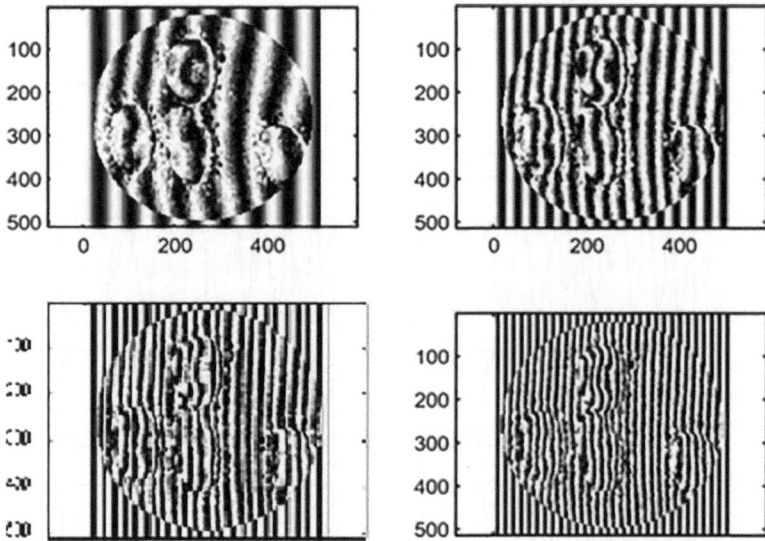

Fig. 8.6 Interferometric images of the coronavirus using the cosine function to represent the phase at four different spatial frequencies: 1/32, 1/64, 1/96, and 1/128. The original image is multiplied by a factor of $\alpha = 1/32$

The map of the refractive index distribution computed from Eq. (8.10) is the final object of this work (Figs. 8.9, 8.10, and 8.11; Tables 8.1, 8.2, 8.3 and 8.4).

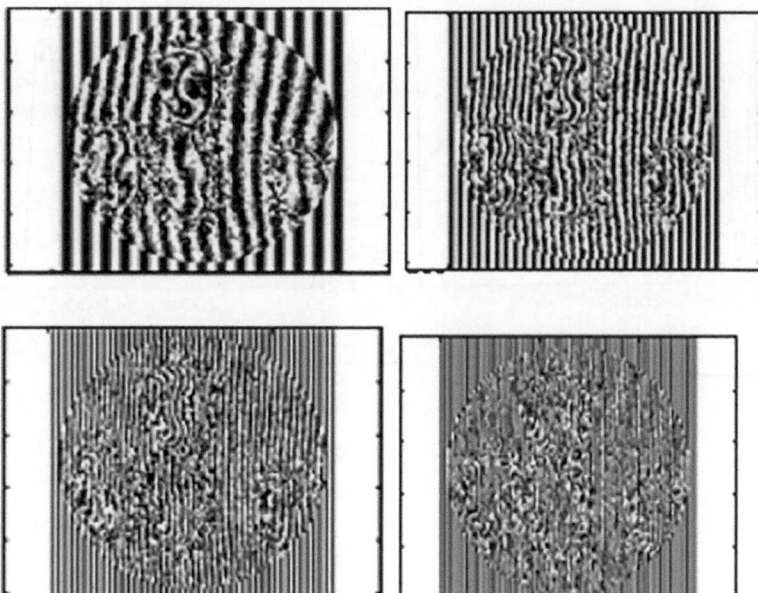

Fig. 8.7 Multiple beam interferometry images of the coronavirus using the Airy function to represent the phase at four different background frequencies, as shown in Fig. 8.6. The original image is multiplied by a factor of $\alpha = 1/32$. The matrix has dimensions of 512×512 pixels

8.4 Conclusions

The phase shift of coronavirus was deduced from the interferometer images. The interferometer images using multiple beam interference provided better contrast than did the corresponding images with two-beam interference, as expected. The effect of the mirror reflection coefficient on multiple beam interference images is discussed (Fig. 8.9).

Useful information obtained from studying this virus using interferometry is extracted from the fringe shift of the modulated interference pattern, namely, the refractive index distribution of the whole image. Consequently, detailed, and precise information about the virus may be extracted from the refractive index distribution. In addition, since the refractive index is directly related to the polychromatic spectral distribution of illuminating light according to the Cauchy formula, the diameter of the virus cell can be accurately observed as it changes with the wavelength of light.

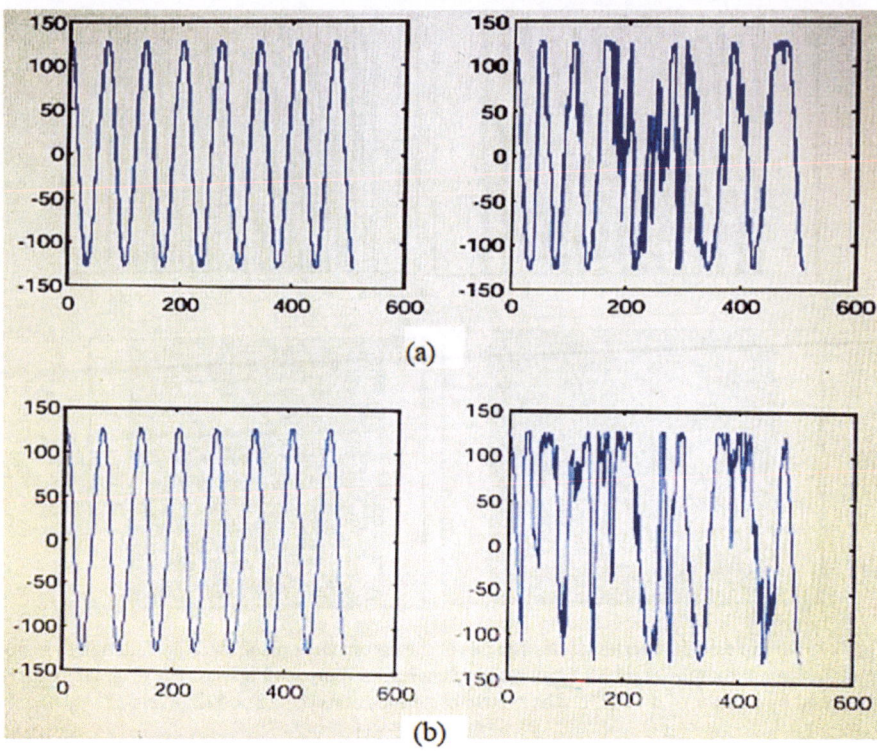

(a)

(b)

Fig. 8.8   a The profile of the coronavirus image taken at constant $x = 150$ pixels extracted from the image shown in Fig. 8.3. Seven fringes are shown; **b** the profile of the coronavirus image taken at constant $x = 330$ pixels extracted from the image shown in Fig. 8.3. Seven fringes are given

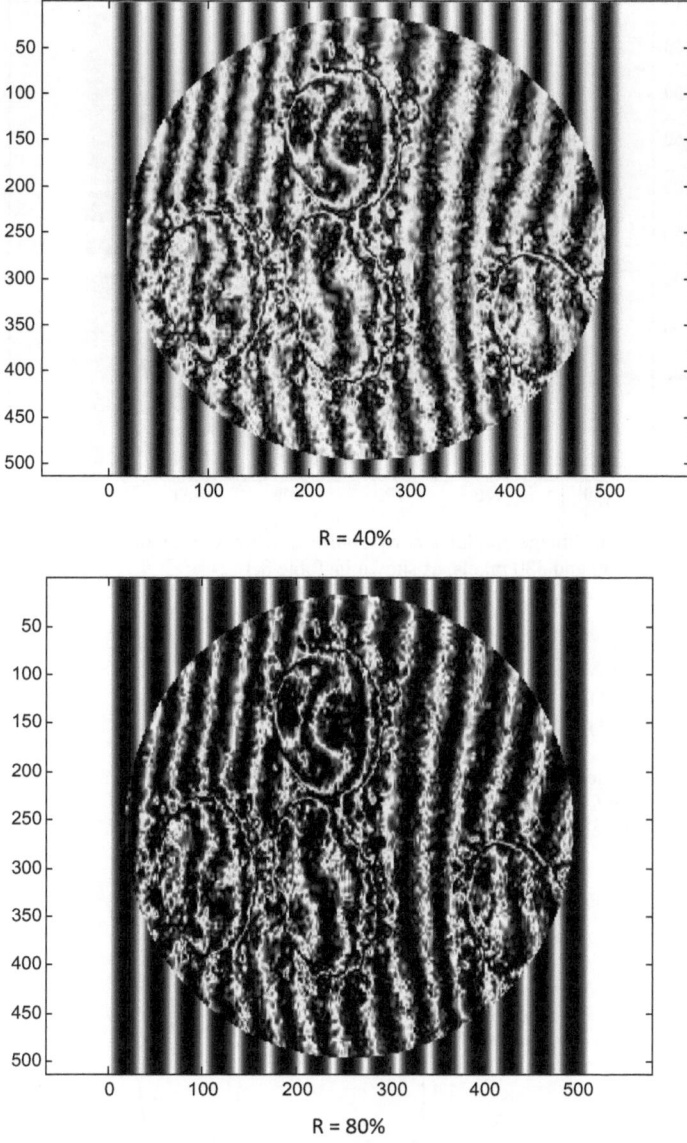

Fig. 8.9 Effect of the mirror reflection coefficient on multiple beam interference images

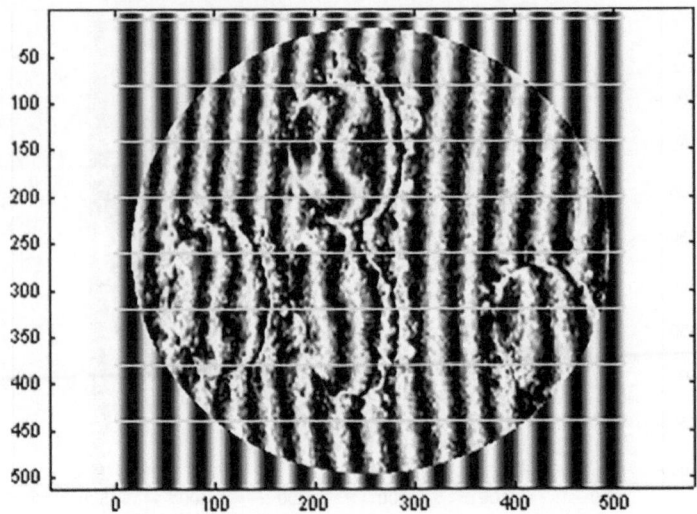

Fig. 8.10 Coronavirus image modulated by fourteen fringes used in the computation of the refractive index at 150 and 330 pixels, as shown in Table 8.1

Fig. 8.11 Plot of the refractive index versus the horizontal distance at the constant line $x = 330$ pixels. Three cells from the coronavirus are scanned in the image. Interfringe spacing $\Delta Z = 34$ pixels

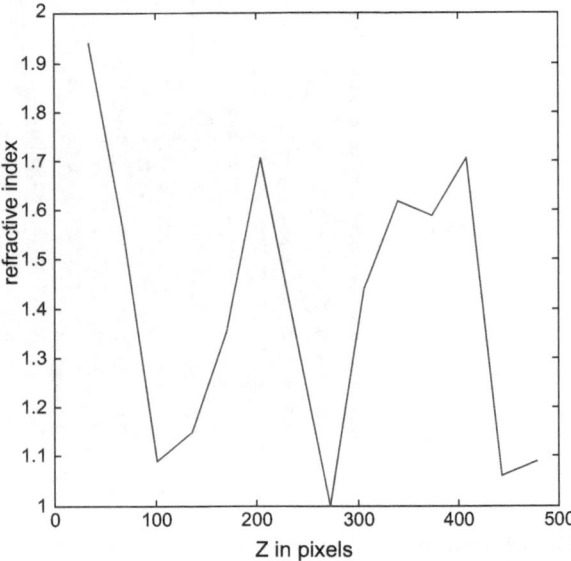

Table 8.1 The refractive index values as a function of the Z coordinate at a certain horizontal line at 330 pixels

Z	Z_{image}	$\delta Z = Z - Z_{image}$	$\mu(Z) = 1 + \delta Z / \Delta Z$
34	2	32	1.94
68	49	19	1.56
102	99	3	1.09
136	131	5	1.15
170	158	12	1.35
204	180	24	1.71
272	272	0	1.00
306	291	15	1.44
340	319	21	1.62
374	354	20	1.59
408	384	24	1.71
443	441	2	1.06
479	476	3	1.09

Table 8.2 The refractive index values as a function of the Z coordinate at a certain horizontal line at 60 pixels

Z	Z_{image}	$\delta Z = Z - Z_{image}$	$\mu(Z) = 1 + \delta Z / \Delta Z$
51	63	12	1.80
69	80	11	1.73
84	96	12	1.80
100	111	11	1.73
118	126	8	1.53
132	141	9	1.60
149	152	3	1.20
166	167	1	1.07

Table 8.3 The refractive index values as a function of the Z coordinate at a certain horizontal line at 80 pixels

Z	Z_{image}	$\delta Z = Z - Z_{image}$	$\mu(Z) = 1 + \delta Z / \Delta Z$
51	62	11	1.73
69	76	7	1.47
84	92	8	1.53
100	108	8	1.53
118	126	8	1.53
132	141	9	1.60
149	153	4	1.27
166	170	4	1.27

Table 8.4 The refractive index values as a function of the Z coordinate at a certain horizontal line at 100 pixels

Z	Z_{image}	$\delta Z = Z - Z_{image}$	$\mu(Z) = 1 + \delta Z / \Delta Z$
51	60	9	1.60
69	75	6	1.40
84	91	7	1.47
100	107	7	1.47
118	128	10	1.67
132	144	12	1.80
149	154	5	1.33
166	171	5	1.33

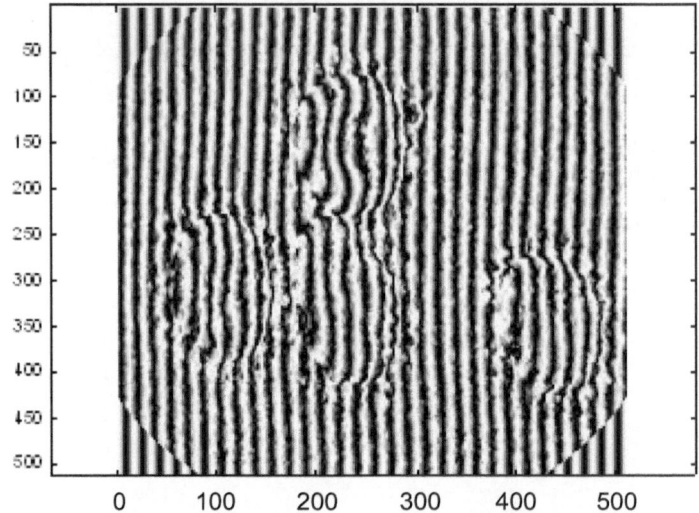

Fig. 8.12 Modulated interference image of the coronavirus, where 32 fringes are shown

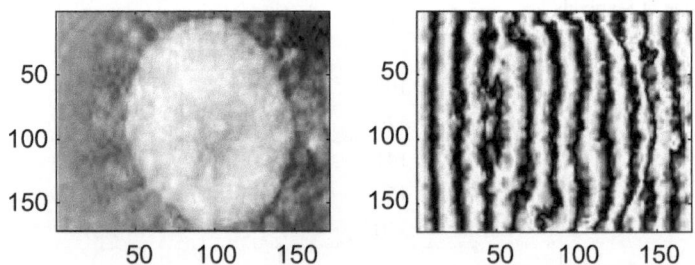

Fig. 8.13 On the left (**a**), a segment from the image shown in Fig. 8.12 is taken at $i = 220{:}390$ pixels and $j = 10{:}180$ pixels. The modulated two-beam interference of the cell (segment) shown in (**b**) is investigated. Both segments in a, and b have dimensions of 170×170 pixels Fig. 8.13

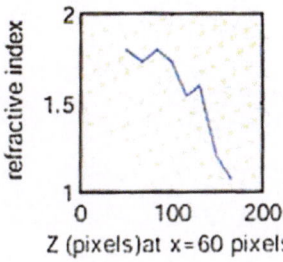 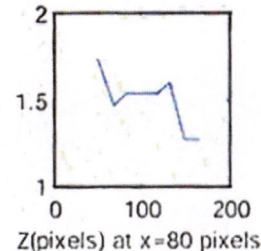 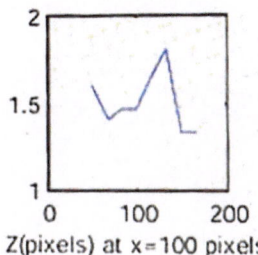

Fig. 8.14 Three plots of the refractive index variation with respect to Z at constant horizontal lines at $x = 60$, 80, and 100 pixels computed from Fig. 8.13b, which has only one cell

References

1. B.W. Neuman, B.D. Adair, J. Virology, Super molecular architecture of severe acute respiratory syndrome corona virus revealed by electron cryo-microscopy. J. Virol. **80**, 7918–7928 (2006). https://doi.org/10.1128/JVI.00645-06
2. J.C. Wyant, Computerized interferometric surface measurements. Appl. Opt. **52**, 1–8 (2013). https://doi.org/10.1364/AO.52.000001
3. R. Crane, Interference phase measurement. Appl. Opt. **8**, 538–542 (1969)
4. J.C. Wyant, Double frequency grating lateral shear interferometer. Appl. Opt. **12**, 2057–2060 (1973). https://doi.org/10.1364/AO.12.002057
5. J.H. Bruning, D.R. Herriott, J.E. Gallagher et al., Digital wave-front measuring interferometer for testing optical surfaces and lenses. Appl. Opt. **13**, 2693–2703 (1974). https://doi.org/10.1364/AO.13.002693
6. J.C. Wyant, Use of an AC heterodyne lateral shear interferometer with real-time wave front correction systems. Appl. Opt. **14**, 2622–2626 (1975). https://doi.org/10.1364/AO.14.002622
7. J.P. Liu, T.C. Poon, Two-step-only quadrature phase shifting digital holography. Opt. Lett. **34**, 250–252 (2009). https://doi.org/10.1364/OL.34.000250
8. T. Yatagai, S. Nakadate, Automatic fringe analysis using digital image processing technique. Opt. Eng. **21**, 432–435 (1982)
9. A.M. Hamed, M.A. Saudy, Image processing of argon glow discharge plasma using interferometry. J. Plasma Phys. **81**, 1–14 (2015). https://doi.org/10.1017/S0022377815000550
10. A.M. Hamed, Numerical speckle images formed by diffusers using modulated conical and linear apertures. J. Mod. Opt. **56**, 1174–1181 (2009). https://doi.org/10.1080/09500340902985379
11. A.M. Hamed, Formation of speckle images formed for diffusers illuminated by modulated apertures (circular obstruction). J. Mod. Opt. **56**, 1633–1642 (2009). https://doi.org/10.1080/09500340903277792
12. A.M. Hamed, Discrimination between speckle images using diffusers modulated by some deformed apertures: simulation. Opt. Eng. **50**, 1–7 (2011). https://doi.org/10.1117/1.3530085
13. A.M. Hamed, Computer generated quadratic and higher order apertures and its application on numerical speckle images. Optics Photon. J. **1**, 43–51 (2011). https://doi.org/10.4236/opj.2011.12007
14. A.M. Hamed, Study of graded index and truncated apertures using speckle images. Precis. Instr. Mechanol. (PIM) **3**, 144–152 (2014)
15. M. Born, E. Wolf, in *Principles of Optics, Electromagnetic Theory of Propagation, Interference and Diffraction of Light*. 2nd Edn (1964), pp. 325

Chapter 9
Investigation of the Colon Using Cascaded Interferometric Techniques

In this chapter, a cascaded interferometric technique is used in the investigation of colon cancer images. The surface topography of colon cancer can be obtained from either the fringe shift corresponding to the interference pattern, or the profile shape plotted from the source input image. In addition, the refractive index corresponding to the colon's malignant and benign cells was computed from the surface height variation as outlined in the analysis. All images and results are obtained by using MATLAB code.

9.1 Introduction

Colon cancer: A malignancy that arises from the inner lining of the colon. Most, if not all, of these cancers develop from colonic polyps. The removal of these precancerous polyps can prevent colon cancer. Colon polyps and early colon cancers cause no signs or symptoms. Full-blown colon cancer can cause occult (a microscopic amount of) blood in the stool, overt rectal bleeding, bowel obstruction, and weight loss. Risk factors for colon cancer include a family history of colon cancer or colonic polyps, and long-standing ulcerative colitis. The overall risk can be reduced by following a diet low in fat and high in fiber. Colon cancer is preventable and curable. It is prevented by removing precancerous colon polyps. It is curable if it is detected early when it can be surgically removed before it has spread to other parts of the body. If screening and surveillance programs are applied universally, there could be a tremendous reduction in the incidence and mortality of colon cancer [1].

An image of a colon with a cancerous zone in the descending colon and a polyp zone in the ascending colon is shown in Fig. 9.1. In the most basic terms, cancer refers to cells that grow out of control and invade other tissues. Cells may become cancerous due to the accumulation of defects, or mutations, in their DNA. Certain inherited genetic defects and infections can increase the risk of cancer. Environmental

A. Hamed, *Cascaded Interferometers and Their Medical Applications*,
SpringerBriefs in Applied Sciences and Technology,
https://doi.org/10.1007/978-3-031-64535-8_9

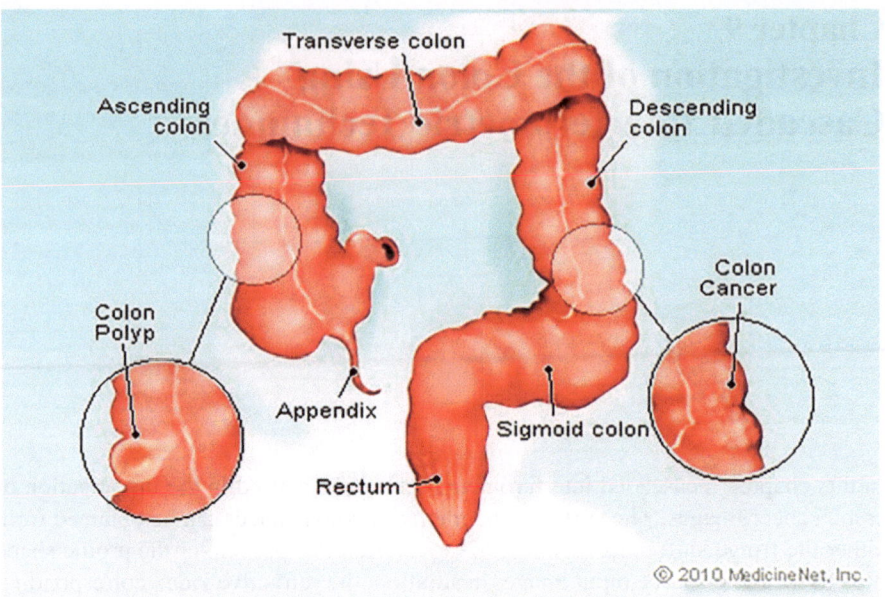

Fig. 9.1 Colons with cancerous cells and polyps are shown in the image

factors (for example, air pollution) and poor lifestyle choices such as smoking and heavy alcohol use can also damage DNA and lead to cancer. Most of the time, cells can detect and repair DNA damage. If a cell is severely damaged and cannot repair itself, it usually undergoes programmed cell death or apoptosis. Cancer occurs when damaged cells grow, divide, and spread abnormally instead of self-destructing as they should.

In the literature, the differentiation between benign and malignant colon cancer is outlined in different articles using segmentation and image processing [2]. Benign tumors grow locally and do not spread. As a result, benign tumors are not considered cancer. They can still be dangerous, especially if they press against vital organs such as the brain. Malignant tumors can spread and invade other tissues. This process, known as metastasis, is a key feature of cancer. There are many different types of malignancies based on where a cancer tumor originates. Images of normal, benign, and malignant colon cells are shown in Fig. 9.2. Recently, the image processing of colon cancer and tumors has been outlined in different publications [3–8]. The symptoms and treatment of various cancers in various parts of the large intestine may differ materially. The main parts of the large intestine are the cecum, ascending colon, transverse colon, descending colon, and rectum. Cancer can occur in any of these parts (mostly in the rectum) [8]. Bleeding from the rectum, observed as the presence of visible blood in the stool, should be a warning about a serious disorder [3, 4]. The analysis algorithm and image processing for the quantitative description of colon cancer cells are described [9]. Later, image processing of corona and SIDA viruses using two and multiple beam interference was described in [10–12], which was

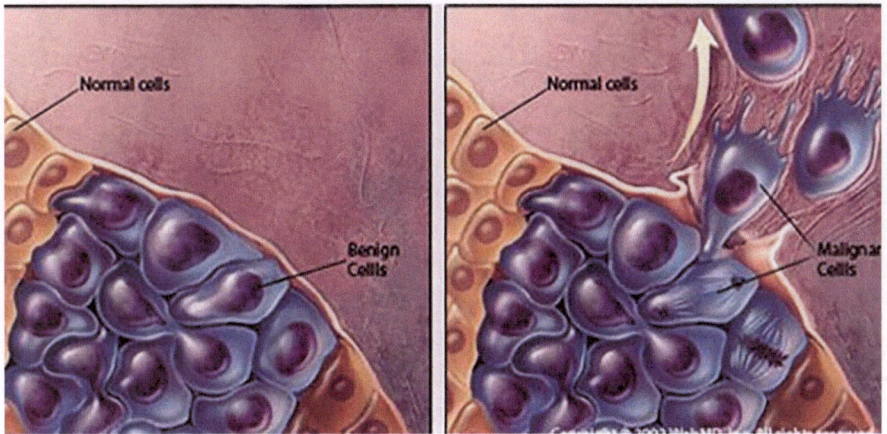

Fig. 9.2 Malignant and benign cells are shown in the colon, whereas benign cells are shown on the left, while malignant cells are shown on the right

improved using cascaded arrangements for interference, as described in [13–16]. In addition, image processing based on the speckle photography technique is described in [10, 17, 18]. In this paper, cancerous cells in the colon were investigated using two different techniques, namely, cascaded interferometry and speckle photography.

9.2 Theoretical Analysis

9.2.1 Cascaded Interferometric Technique

The intensity of the two-beam interference obtained in the detector plane can be expressed as the modulus squared as follows:

$$I(x, y; z) = I_0\{1 + M \ \cos[\Phi(x, y; z) - \psi(x, y)]\} \tag{9.1}$$

where $I(x, y; z)$ is the intensity of the modulated interference field at point $I\ (x, y)$ for object depth z, $I_0 = |R|^2 + |a|^2$, is the function that characterizes the mean intensity of the interference pattern, and $M = (2a\ R)/[|R|^2 + |a|^2]$, is the function that determines the modulation of the interference signal. In this case, the obtained trigonometric function has straight-line fringes modulated by the object phase information. The modulated intensity is rewritten in matrix form as follows:

$$I(x, y; z) = I_0 \sum_{n=1}^{N} \sum_{m=1}^{M} \{1 + M \ \cos[\Phi(n\Delta x, m\Delta y; z) - \psi(n\Delta x, m\Delta y)]\} \tag{9.2}$$

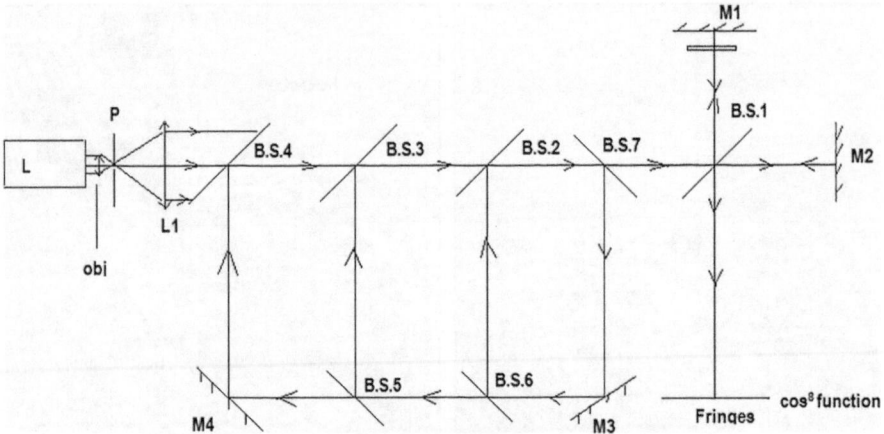

Higher Order two beam interference arrangement of multiples of cos² function.

Fig. 9.3 A cascaded higher-order two-beam interference where three loops are added to the original arrangement, giving the $\cos^8$ function.

The coherent multiplication of two-beam interference is governed by the feedback of the coherent laser beam incident on the Michelson arrangement, as shown in Fig. 9.3. Hence, the number of feedback passes (N) on the interferometer gives an intensity distribution in the form [13]:

$$I_{\text{feedback}}(x, y; N) = I_0 \cos^{2(N+1)}(\delta) \tag{9.3}$$

The ordinary two-beam interference has an intensity distribution extracted from Eq. 9.3, where $N = 0$ (no feedback) as follows:

$$I_{\text{ordinary}}(x, y; 0) = I_0 \cos^2(\delta) \tag{9.4}$$

For a single feedback, $N = 1$

$$I_{\text{feedback}}(x, y; N = 1) = I_0 \cos^4(\delta) \tag{9.5}$$

For two feedback light passes, $N = 2$

$$I_{\text{feedback}}(x, y; N = 2) = I_0 \cos^6(\delta) \tag{9.6}$$

It is known that each feedback pass adds $\cos^2\delta$ to the original $\cos^2\delta$.

The fringe sharpness is conveniently measured by the full intensity width at half maximum (FWHM) [14]. In the case of ordinary two-beam interference, the fringe sharpness is computed as:

$$\frac{I}{I_0} = \frac{1}{2} = \cos^2(\delta_w) \tag{9.7}$$

In this case, the FWHM represented by (δ_w) is computed as follows:

$$\delta_w = \cos^{-1}\left(\frac{1}{\sqrt{2}}\right) \tag{9.8}$$

In the case of multiple feedback of the number of passes N, the FWHM is computed as follows:

$$\delta_w = \cos^{-1}\left(\frac{1}{\left[\sqrt{2}\right]^{\frac{1}{N+1}}}\right) = \cos^{-1}\left\{\frac{1}{(2)^{\frac{1}{2(N+1)}}}\right\} \tag{9.9}$$

In the case of cascaded multiple beam interference, as shown in Fig. 9.4, the intensity distribution is the Airy distribution raised to a power of N, where N is the number of cascaded interferometers. Then, the intensity distribution is represented as follows [15, 16]:

$$I(\delta; R, N) = \left[\frac{T^2}{1 + R^2 - 2R\cos(\delta)}\right]^N \tag{9.10}$$

The maximum intensity is computed as:

$$I(\delta = 2\pi; R, N) = I_{\max}(R, N) = \left[\frac{T^2}{(1 - R)^2}\right]^N \tag{9.11}$$

Using Eqs. 9.10 and 9.11, the normalized intensity due to cascaded multiple beam interference can be written as follows:

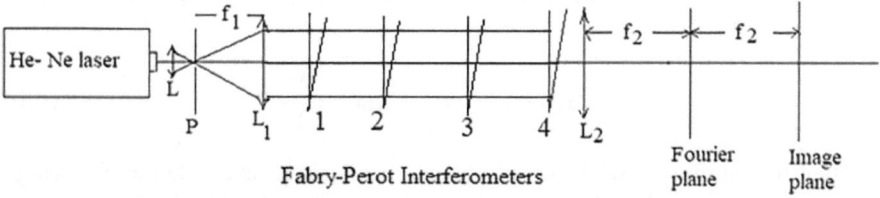

Fig. 9.4 Multiple beam interferometer composed of four cascaded interferometers. L objective lens, P pinhole, and L_1 converging lens where the elements L, P, and L_1 render the laser beam spatially filtered and collimated. L_2 is the Fourier transform lens of focal length f_2. The Fourier and imaging planes are shown in the figure

$$I_{\text{normalized}}(\delta; R, N) = \frac{I}{I_{\text{max}}} = \frac{1}{\left[1 + F \, \sin^2\left(\frac{\delta}{2}\right)\right]^N} \tag{9.12}$$

where

$$F = \frac{4R}{(1 - R)^2}, \text{ is the coefficient of finesse.} \tag{9.13}$$

F is a measure of fringe sharpness and contrast.

9.2.2   Computation of the Refractive Index from the Interference Image

Once the phase is determined across the interference field, the corresponding height distribution $h\,(x, y)$ on the surface of the coronavirus can be determined [14] as follows:

$$h(x, y; z) = \frac{\lambda}{4\pi} \Phi(n\Delta x, m\Delta y; z) \tag{9.14}$$

We assumed that the surface was measured at normal incidence. Almost all interferometers used to measure surface height variations use phase-shifting techniques.

The optical path difference represents the height variation of the image, namely, $h\,(x, y, z)$; then, the refractive index becomes [14]:

$$\mu(x, y) = 1 + \frac{\mathrm{d}}{\mathrm{d}z}\left[h(x, y, z)\right] \tag{9.15}$$

The differentiation of the height distribution $h\,(x, y, z)$ for z gives the differential fringe shift and the amplitude of the planar image a (x, y). Consequently, we finally obtain Eq. 9.16.

$$\mu(x, y)_{\text{const}.x} = 1 + a(x, y)\frac{\delta z}{\Delta z} \tag{9.16}$$

The fringe shift is δz, while the interfringe spacing is Δz at constant x. The fringes are assumed to be in the x–y plane, where z is the axis normal to the fringe system, which represents the height depth, and a (x, y) represents the amplitude of the image.

In Eq. 9.16, $h\,(x, y, z) = a\,(x, y)\,\delta z$.

9.3 Results and Discussion

The contour of the colon cancer cells was detected using two-beam interference, as shown in Fig. 9.7a, corresponding to the source image in Figs. 9.5 and 9.6. It is improved by using cascaded multiple beam interference and compared with ordinary interference patterns, as shown in Fig. 9.7b, c. Benign and normal cells in the human colon modulated with two beams interference corresponding to the source image, as shown in Fig. 9.6a, are shown in Fig. 9.8a. The malignant cells modulated with the same ordinary two-beam interference are shown in Fig. 9.8b, corresponding to the image in Fig. 9.6b.

In the 1st row, Fig. 9.9a, a benign cell of the colon is shown, in the 2nd row, the corresponding modulated fringe pattern originated from ordinary two-beam interference; and in the 3rd row, the profile plot of the benign fringe pattern at 256 pixels and in the 4th-row plot of ordinary two-beam interference are given. Figure 9.9b shows a malignant cell of the colon in the middle of the corresponding modulated fringe pattern, while the lower profile plot corresponds to the malignant fringe pattern at 256 pixels. The improved sharp fringe pattern attained, as shown in Figs. 9.9a and 9.10a, using multiple beam interference corresponding to the benign and malignant cells is further improved using cascaded interference, as shown in Figs. 9.8c and 9.11c. The corresponding plots are shown in Figs. 9.10b and 9.11b for the ordinary pattern and in Figs. 9.10d and 9.11d for the cascaded interference pattern.

The fringe shift corresponding to the benign and malignant cells is accurately computed from the sharp profiles obtained in the case of cascaded multiple beam interference, as shown in Fig. 9.10d for the benign cell and Fig. 9.11d for the malignant cell. The fringe sharpness is proportional to the number of cascaded passes N.

The height of the surface topography corresponding to the benign and malignant colon cells is extracted from the fringe shift for a certain amplitude a (x, y) in the

Fig. 9.5 The source image of the colon cancer cells

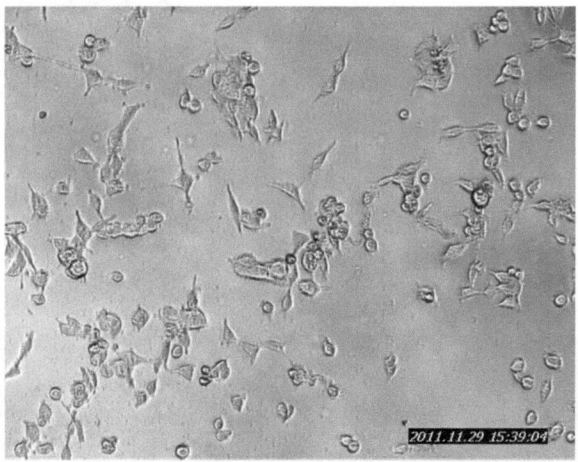

Fig. 9.6 a Benign and
normal cells are visible in the
image of the colon.
b Malignant and normal cells
are shown in the image of the
colon

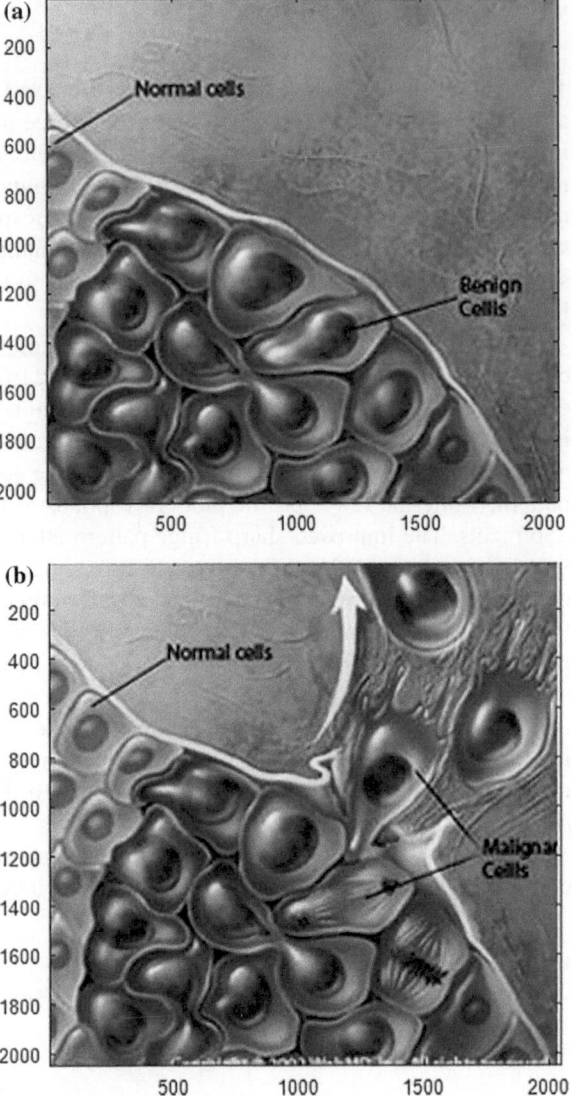

image, as shown in Tables 9.1 and 9.2. In addition, the corresponding refractive index
is computed using Eq. 9.16.

Fig. 9.7 **a** Modulated cancerous cells using ordinary two-beam interference. The contour corresponding to the cancerous cells in the colon corresponding to the source image shown in Fig. 9.5 is shown. **b** Modulated cancerous cells using ordinary multiple beam interference. **c** Modulated cancerous cells using cascaded multiple beam interference with $N = 5$

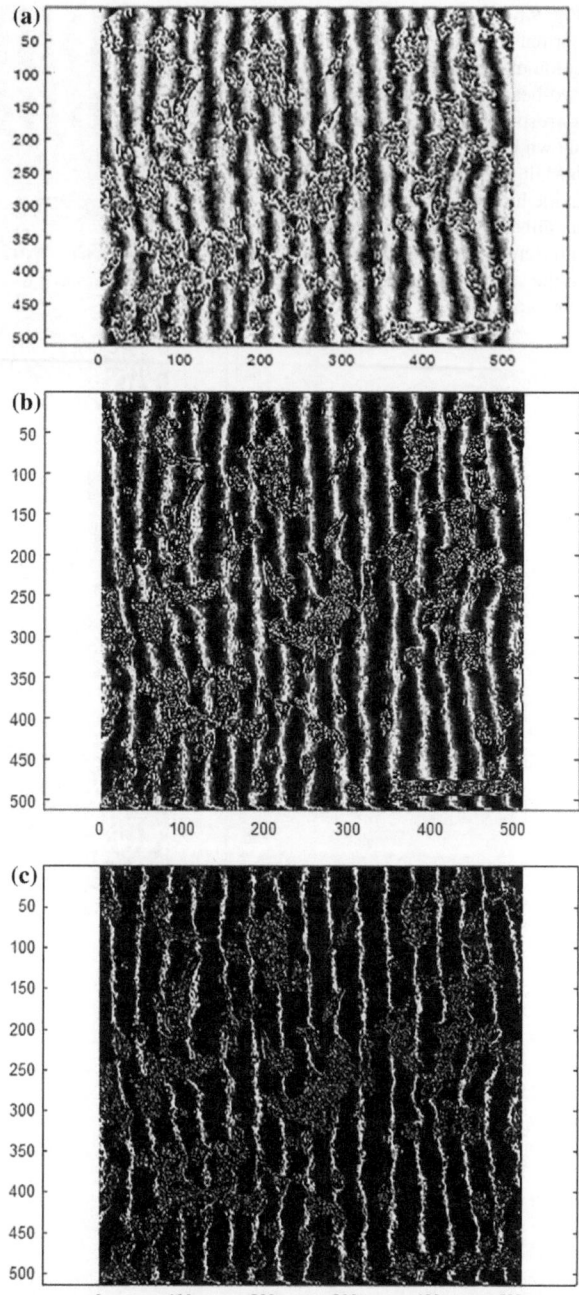

Fig. 9.8 **a** Benign and normal cells in the human colon modulated with two-beam interferences corresponding to the image shown in Fig. 9.6a. **b** Malignant and normal cells in the human colon modulated with two-beam interferences corresponding to the source image shown in Fig. 9.6b

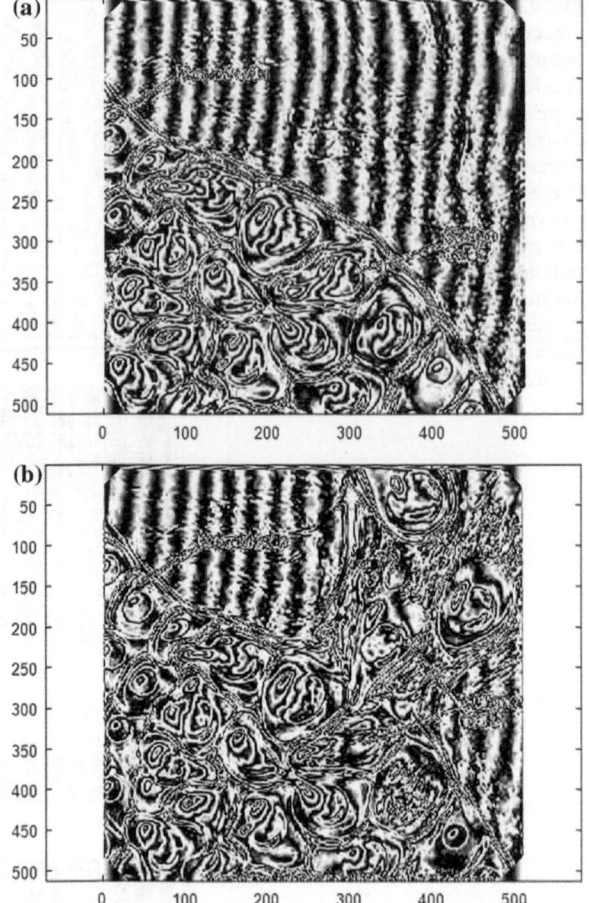

(a)

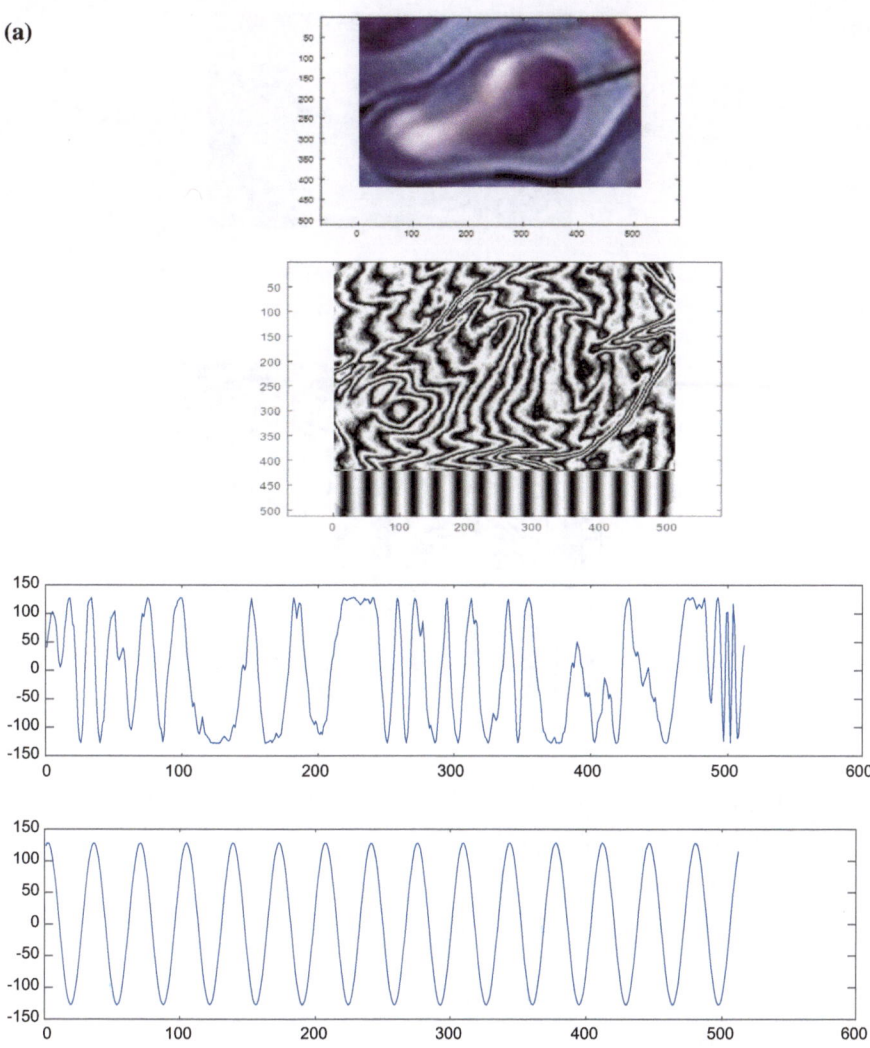

Fig. 9.9 a In the 1st row, a benign cell of the colon is shown; in the 2nd row, the corresponding modulated fringe pattern is shown; in the 3rd row, the profile plot of the benign fringe pattern at 256 pixels is shown; and in the 4th row, the plot of ordinary two-beam interference is shown. **b** A malignant cell of the colon is shown above, in the middle of the corresponding modulated fringe pattern, while it is shown in the lower profile plot corresponding to the malignant fringe pattern at 256 pixels

(b)

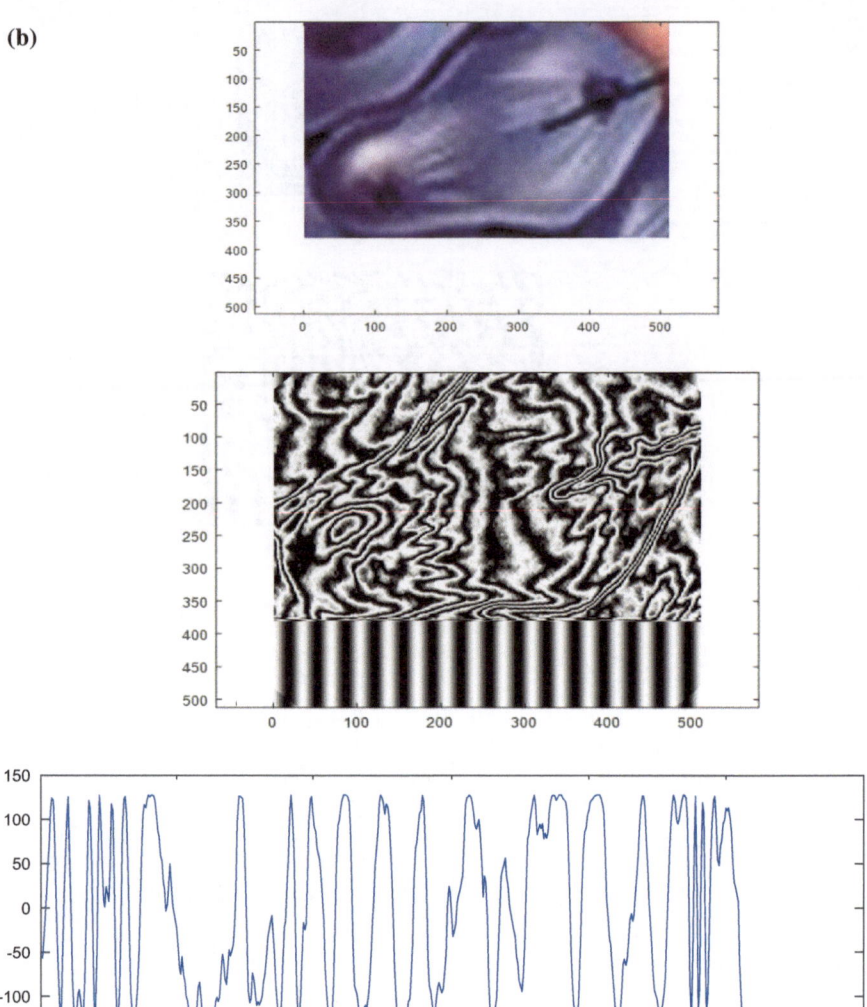

Fig. 9.9 (continued)

9.4   Conclusion

We used ordinary two-beam interference and multiple beam interference for discrim-
inating between benign and malignant colon cells. The fringe shift differed between
the two groups. Further improvement in the sharpness of the fringe shift is attained
by using cascaded models of two and multiple beam interference. It was shown that
the fringe sharpness improves with the number of cascaded passes N [13–16]. This

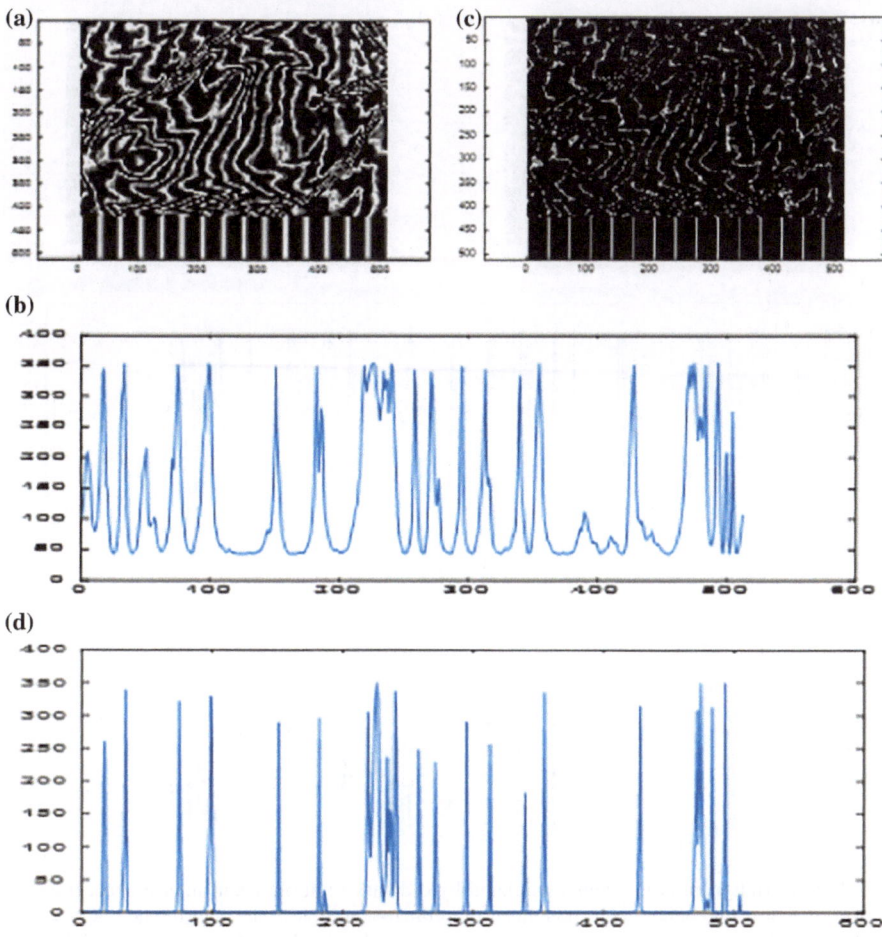

Fig. 9.10 **a** Benign colon cell modulated by ordinary multiple beam interference. **b** Profile corresponding to the benign cell modulated by ordinary multiple beam interference of $N = 1$ at the horizontal section at 256 pixels. **c** Benign colon cell modulated by cascaded multiple beam interference of $N = 10$. **d** Profile corresponding to the benign cell modulated by cascaded multiple beam interference of $N = 10$ at the horizontal section at 256 pixels

approach is useful for investigating fringe shifts because it provides more accurate results.

The height of the surface topography corresponding to the benign and malignant colon cells is extracted from the fringe shift for a certain amplitude a (x, y) in the image. In addition, the corresponding refractive index is computed using the approximate equation obtained in the analysis.

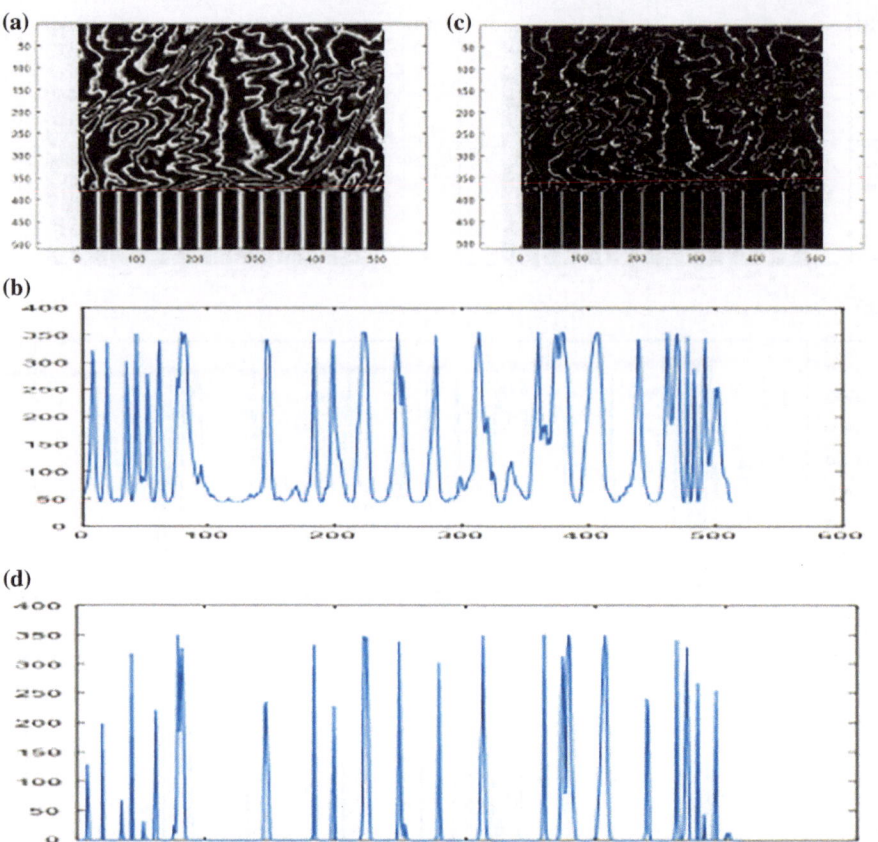

Fig. 9.11 a Malignant colon cells modulated by ordinary multiple beam interference. **b** Profile corresponding to the benign cell modulated by ordinary multiple beam interference of $N = 1$ at the horizontal section at 256 pixels. **c** Malignant colon cell modulated by cascaded multiple beam interference of $N = 10$. **d** Profile corresponding to the malignant cell modulated by cascaded multiple beam interference of $N = 10$ at a horizontal section at 256 pixels

Table 9.1 Benign cell interference results at the horizontal line at 200 pixels; the interfringe spacing $\Delta Z = 34$ pixels compared with the fringe shift shown in the plots

Z	Z_{image}	$\delta Z = Z - Z_{image}$	$\mu = 1 + \delta Z / \Delta Z$
34	18	16	1.4700
68	34	34	2.0000
102	75	28	1.8235
136	102	34	2.0000
170	151	19	1.5588
204	182	22	1.6470
238	234	4	1.1176
272	258	14	1.4117
306	272	34	2.0000

Table 9.2 Malignant cell interference results at the horizontal line at 200 pixels; the interfringe spacing $\Delta Z =$ 34 pixels compared with the fringe shift shown in the plots

Z	Z_{image}	$\delta Z = Z - Z_{image}$	$\mu = 1 + \delta Z / \Delta Z$
34	20	14	1.4117
68	48	20	1.5882
102	82	20	1.5882
136	117	19	1.5588
170	147	23	1.6765
204	198	6	1.1765
238	222	16	1.4700
272	249	23	1.6765
306	279	27	1.7941

References

1. Medical Net. Inc. (2010)
2. N. Jawad, N. Direkze, S.J. Leedham, Inflammatory bowel disease and colon cancer. Rec. Res. Cancer Res. **185**, 99–105 (2011)
3. B.A. Macaskill, P. Chan et al., bowel cancer symptoms do not indicate colorectal cancer and polyps: a systematic review. BMC Gastroenterol. **11**(65) (2011). https://doi.org/10.1186/1471-230X-11-65
4. M. Astin, T. Griffin et al., The diagnostic value of symptoms for colorectal cancer in primary care: a systematic review. Br. J. General Pract. J. Royal Coll. General Practition. **61**(586), 231–243 (2011). https://doi.org/10.3399/bjgp11X572427
5. D. Cunningham, W. Atkin et al., Colorectal cancer. Lancet **375**(9719), 1030–1047 (2010). https://doi.org/10.1016/S0140-6736(10)60353-4
6. D.L. Diggsa, J.N. Myersa et al., Influence of dietary fat type on benzo(a)pyrene [B(a)P] biotransformation in a B(a) P-induced mouse model of colon cancer. J. Nutr. Biochem. **24**(12), 2051–2063 (2013)
7. L.H. Eadie, C.B. Reid, A.J. Fitzgerald, V.P. Wallace, Optimizing multi-dimensional terahertz imaging analysis for colon cancer diagnosis. Expert Syst. Appl. **40**(6), 2043–2050 (2013)
8. A. Bochenek, M. Reicher, Anatomia Człowieka. **2**, Wydawnictwo Lekarskie PZWL, pp. 239–267 (2004)
9. A. Wójcicka1 et al., Advances in intelligent systems and computing **283** (2014) ISBN:978–3–319–06592–2
10. A.M. Hamed, Investigation of SIDA Virus (HIV) images using interferometry and speckle techniques. Int. J. Innov. Res. Comput. Sci. Tech (IJIRCST) **4**, 38–45 (2016)
11. A.M. Hamed, Image processing of coronavirus using interferometry. Opt. Photon. J. **6**, 75–86 (2016)
12. A.M. Hamed, T.A. Al-Saeed, Reconstruction of the corneal layers affected by a periodic noise application on microscopic interferometry. Int. J. Photon. Optic. Technol. (IJPOT) **2**, 6–12 (2016)
13. A.M. Hamed, A modified Michelson interferometer and an application on microscopic imaging. Int. J. Photon. Optic. Technol. (IJPOT) **3** (2017)
14. A.M. Hamed, Processing of the retinal artery image using higher orders of two-beam interference. Int. J. Photon. Optic. Technol. (IJPOT) **3** (2017)
15. A.M. Hamed, Sharp fringes using cascaded Multiple beam interferometers. Appl. Kidney Images Int. J. Innov. Res. Eng. Manag. (IJIREM) **5**, 189–195 (2018)
16. A.M. Hamed, Recognition of some modulated apertures using the Cascaded Fabry-Perot Interferometer (CFPI). Int. J. Innov. Res. Eng. Manag. (IJIREM) **5**, 173–181 (2018)

17. A.M. Hamed, Discrimination between normal and diseased stomach using speckle imaging. Int. J. Innov. Res. Eng. Manag. (IJIREM) **3**, 125–133 (2016)
18. A.M. Hamed, Discrimination between microscopy images and their reconstruction using intelligent digital speckle images. Int. J. Adv. Res. Electron. Commun. Eng. (IJARECE) **8**, 1–8 (2016)